Hello, happy to see you here.

I'm BB, aka Bead Baby.

Abacus is fun, if you know All About it.

"How to Abacus" has books to help you.

Book 1: All About Abacus

Book 2: All About Formulas

Book 3: All About Multiplication

Book 4: All About Division

I will definitely be with you through all the lessons.

Let's go!

To those interested in Abacus

~Sheena

For anyone willing to engage in abacus learning

~Hanlong

For information address PinGrow Media, 1049 El Monte Avenue Ste C # 586, Mountain View California 94040 or contact@pingrow.com

ISBN-13: 978-1-949622-00-3

Visit www.pingrow.com

How to Abacus 1
All About Abacus

Chapter 1
Abacus Now and Then:

The Abacus that this book focuses on is Chinese Abacus though there are abacuses from various regions.

In ancient China, the Abacus was the tool used to help people calculate numbers, as shown in Picture 1. This kind of Abacus can still be found in some stores.

Picture 1

Abacus has been around for thousands of years and now evolved, as shown in Picture 2. The modern type of Abacus is most commonly used and can be purchased through internet or in local Asian stores.

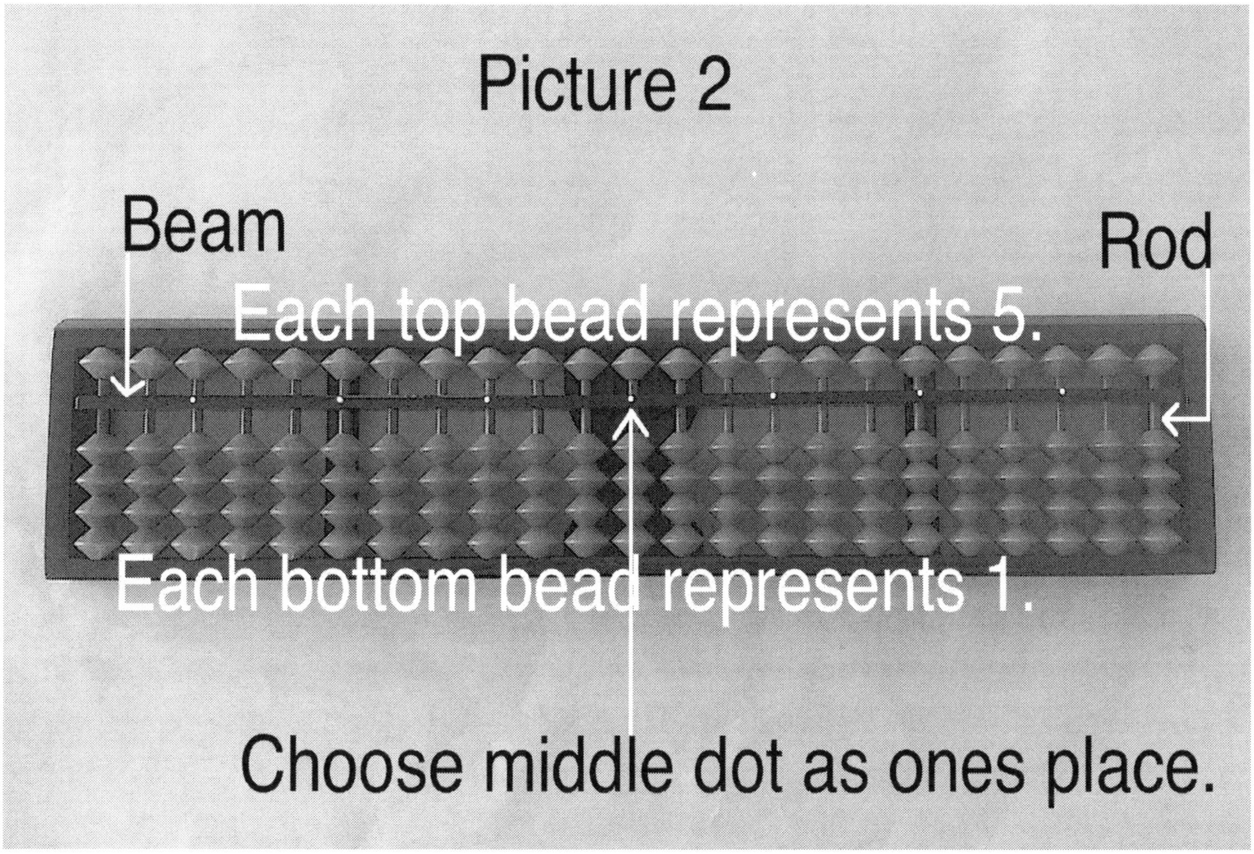

Picture 2 shows beam, dots, beads and rods. Each dot on the beam represents an interval of four place values. Usually, the dot on the center column represents the ones place. The beam divides the beads with the value of five on top and one on the bottom respectively.

Chapter 2

Numbers on Abacus:

What does it look like when the value is 0, 1, 2, 3, 4, 5, 6, 7, 8, 9, 10 and 11?

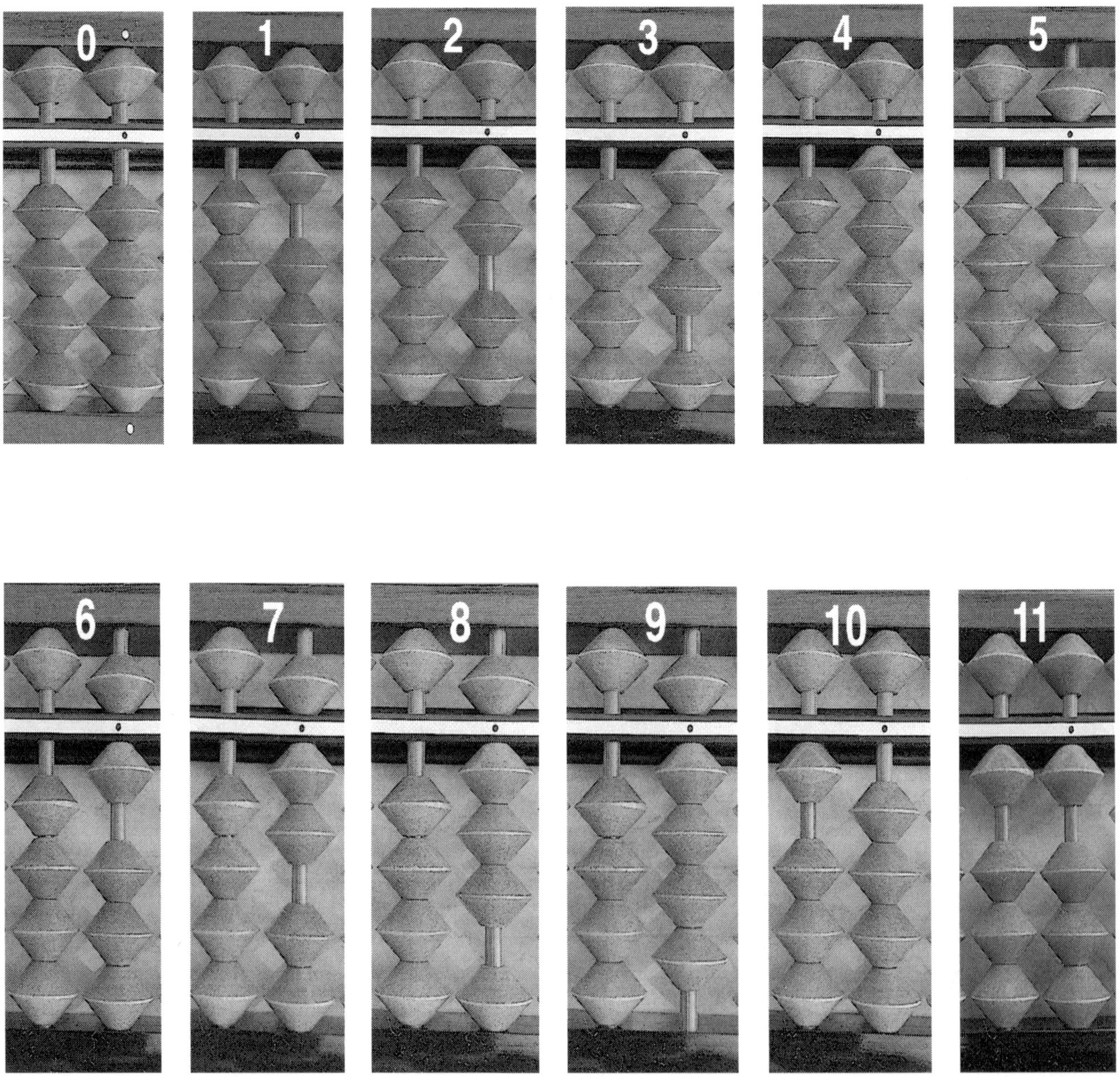

Place Value on Abacus

There is a dot at every 4 rods.
Choose the dot in the middle of the Abacus to be the ones place for 0, 1, 2, 3, 4, 5, 6, 7, 8 and 9. Tens place is on the left side of ones place. Hundreds place is on the left side of tens place. Thousands place is on the left side of hundreds place and so on.

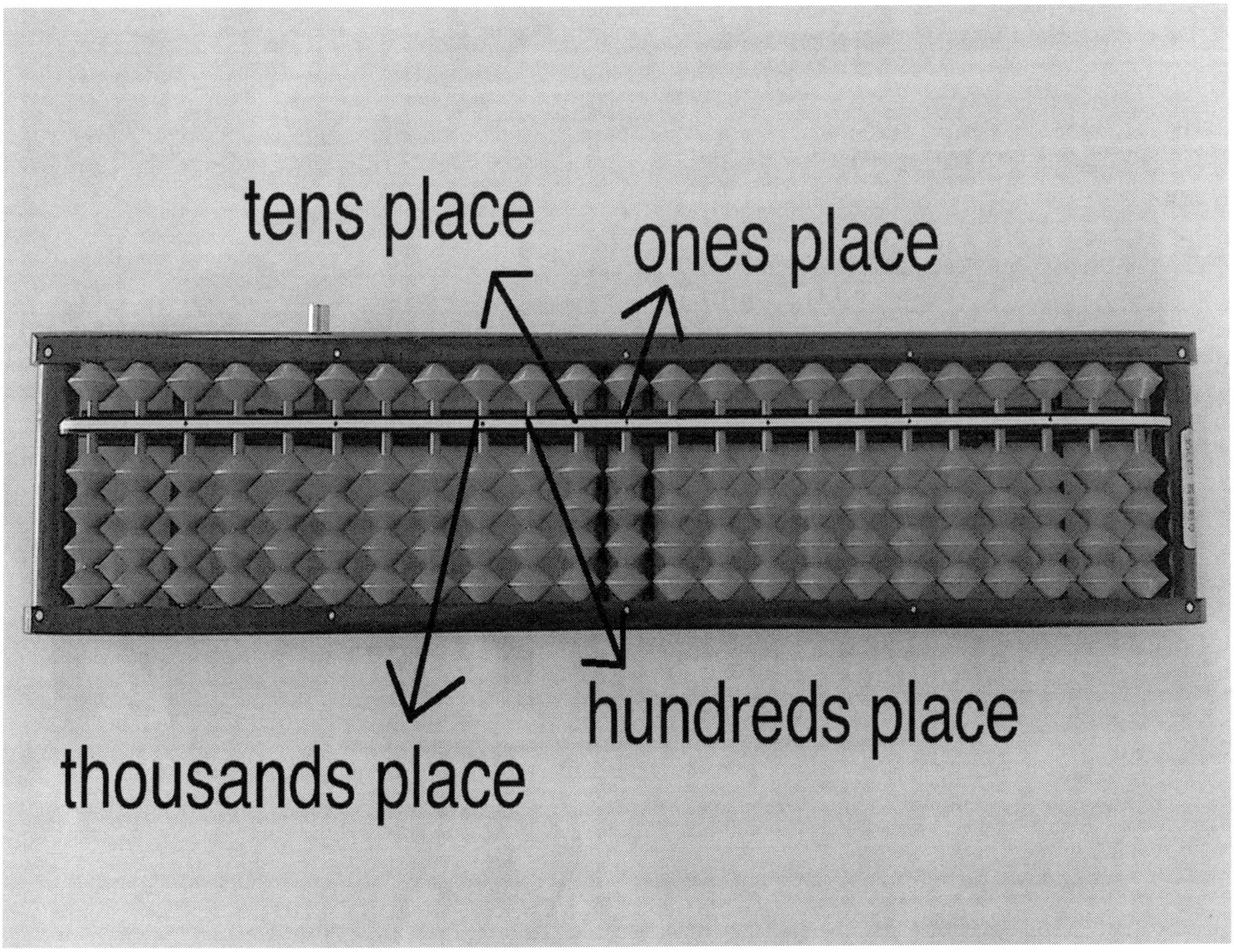

Exercises: Abacus to Numbers

Directions: write down the value of the beads you see. (answers on p.44)

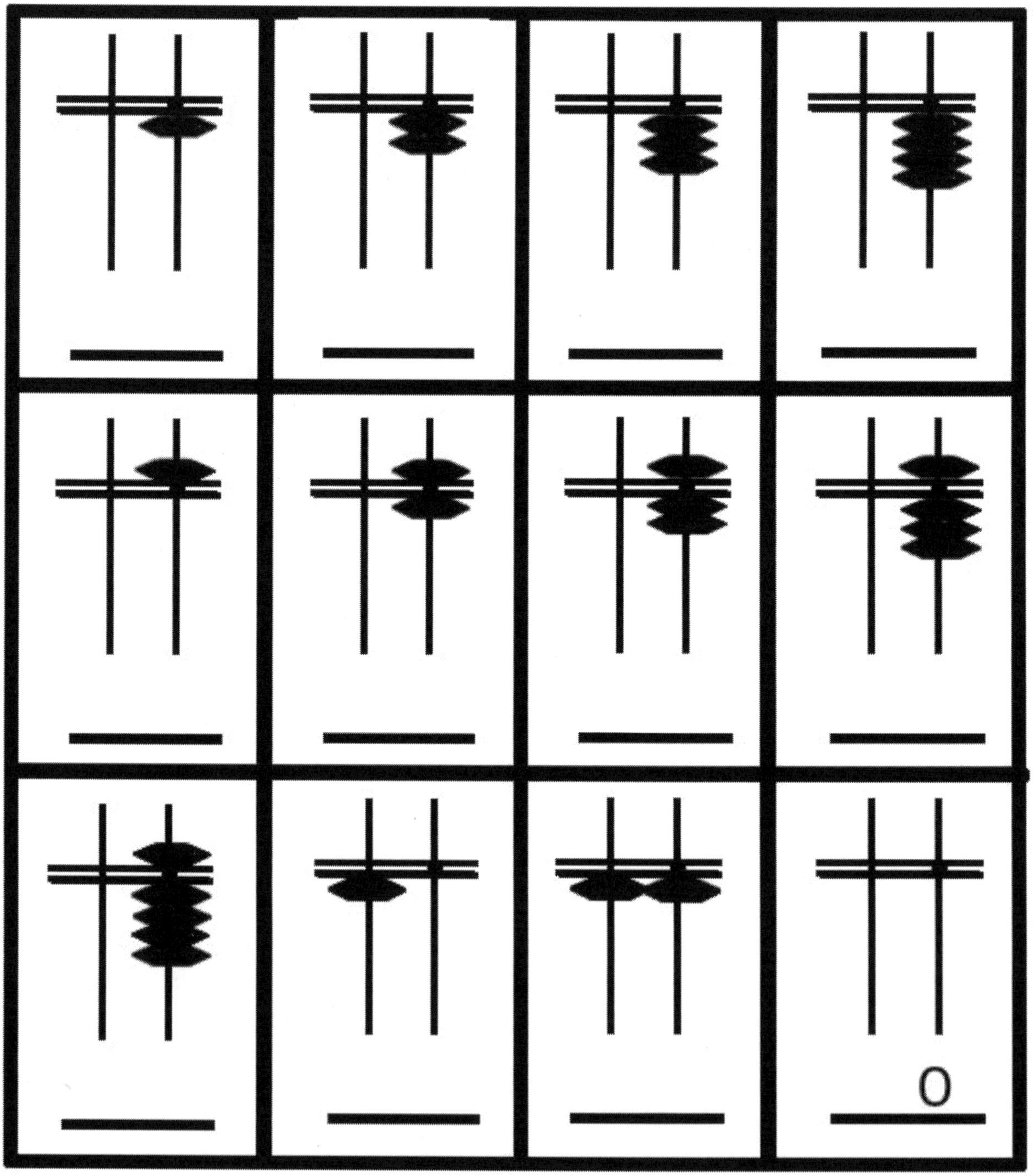

Exercises: Abacus to Numbers

Directions: write down the value of the beads you see. (answers on p.44)

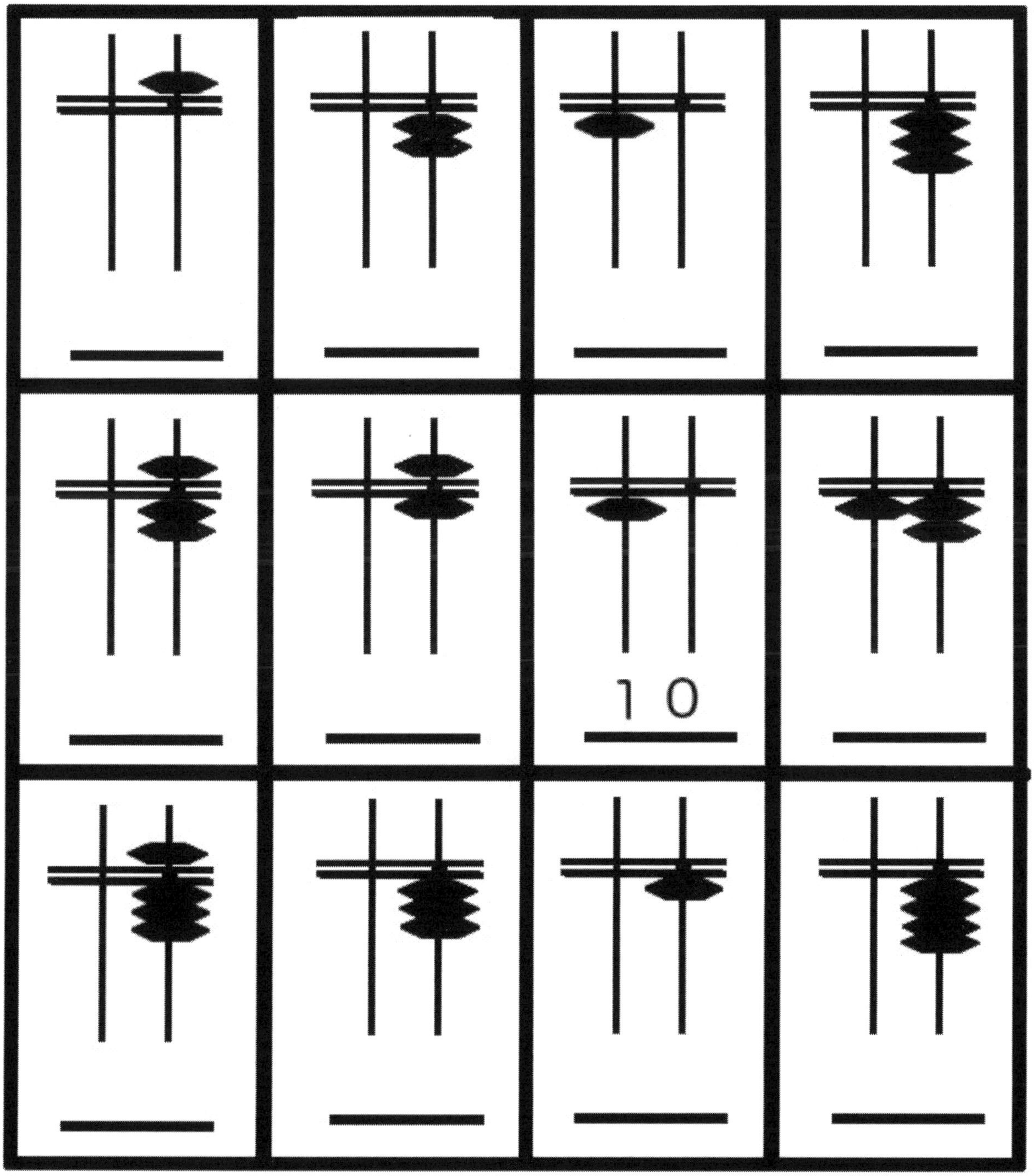

Exercises: Abacus to Numbers

Directions: write down the value of the beads you see. (answers on p.44)

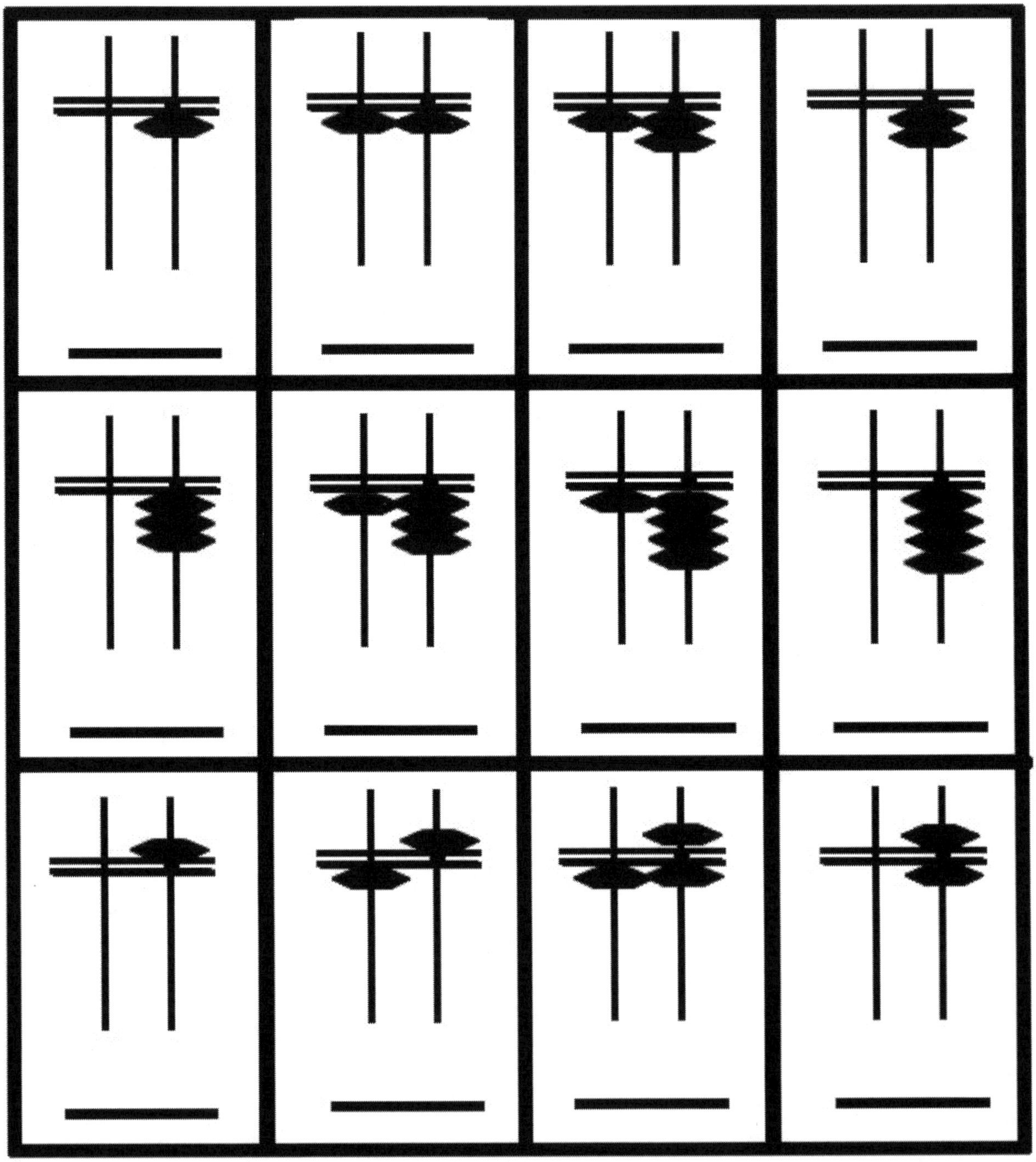

Exercises: Abacus to Numbers

Directions: write down the value of the beads you see. (answers on p.44)

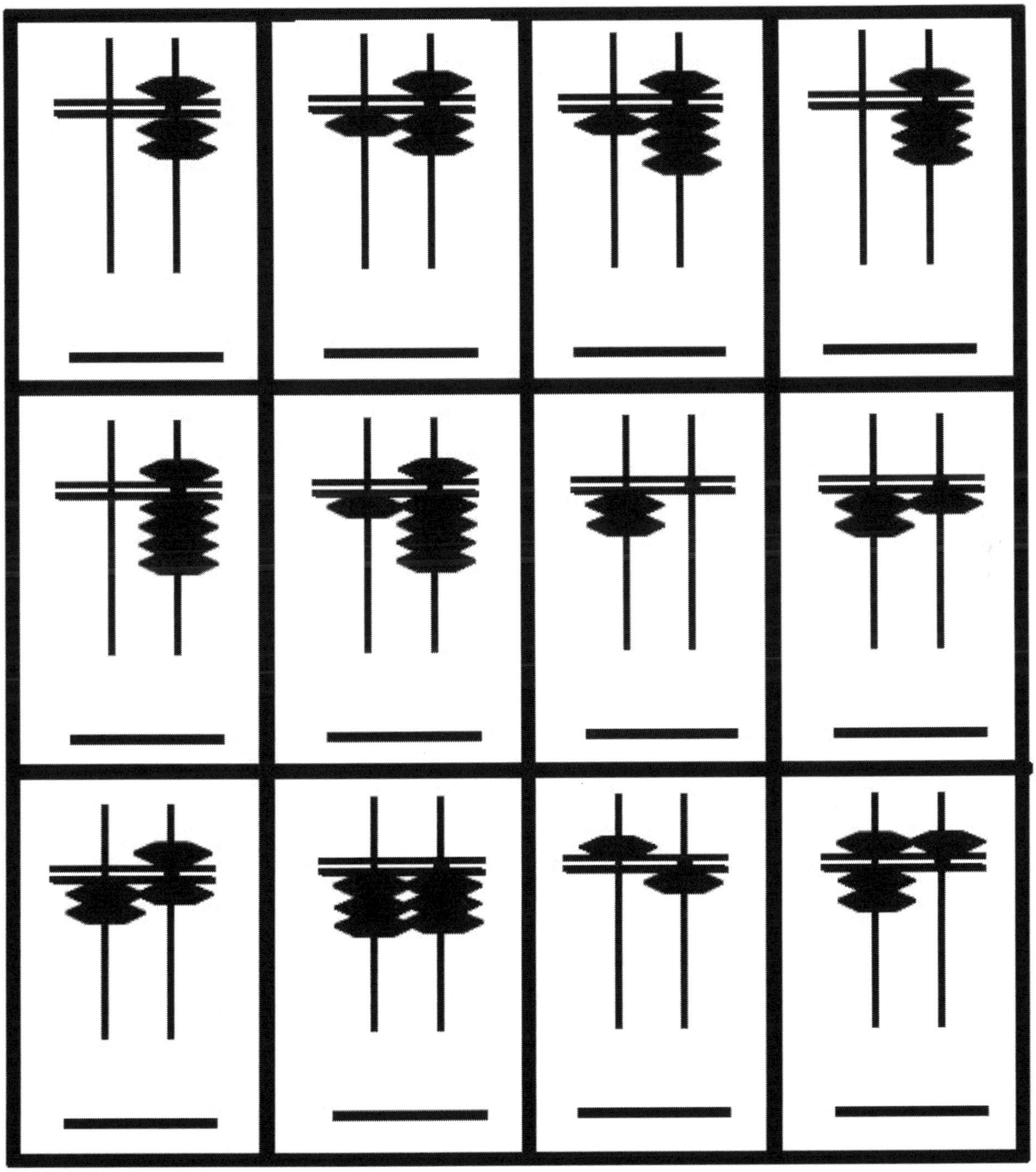

Exercises: Numbers to Abacus

Directions: Draw in the beads that matches the numbers. (answers on p.44)

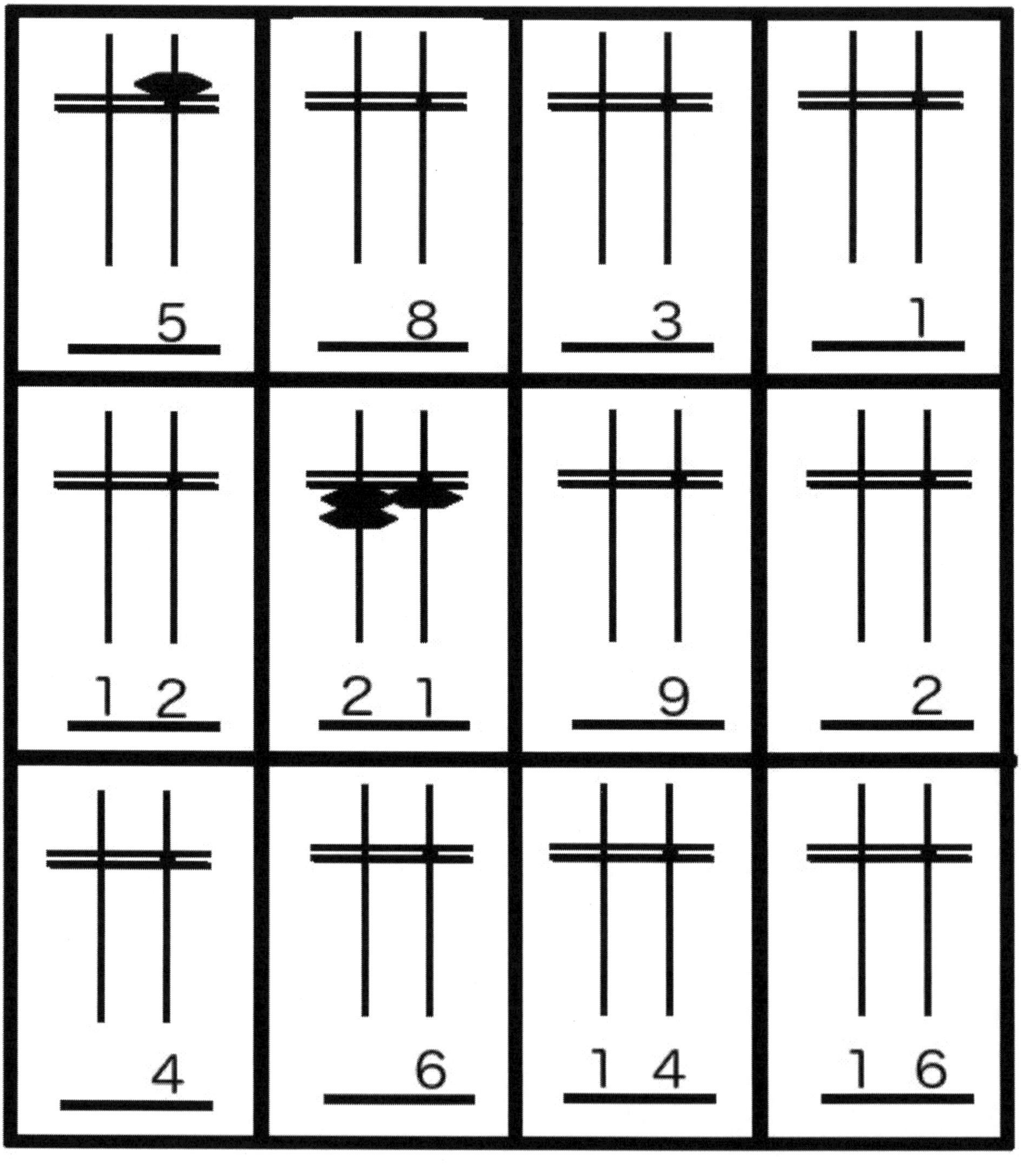

Exercises: Numbers to Abacus

Directions: Draw in the beads that matches the numbers. (answers on p.44)

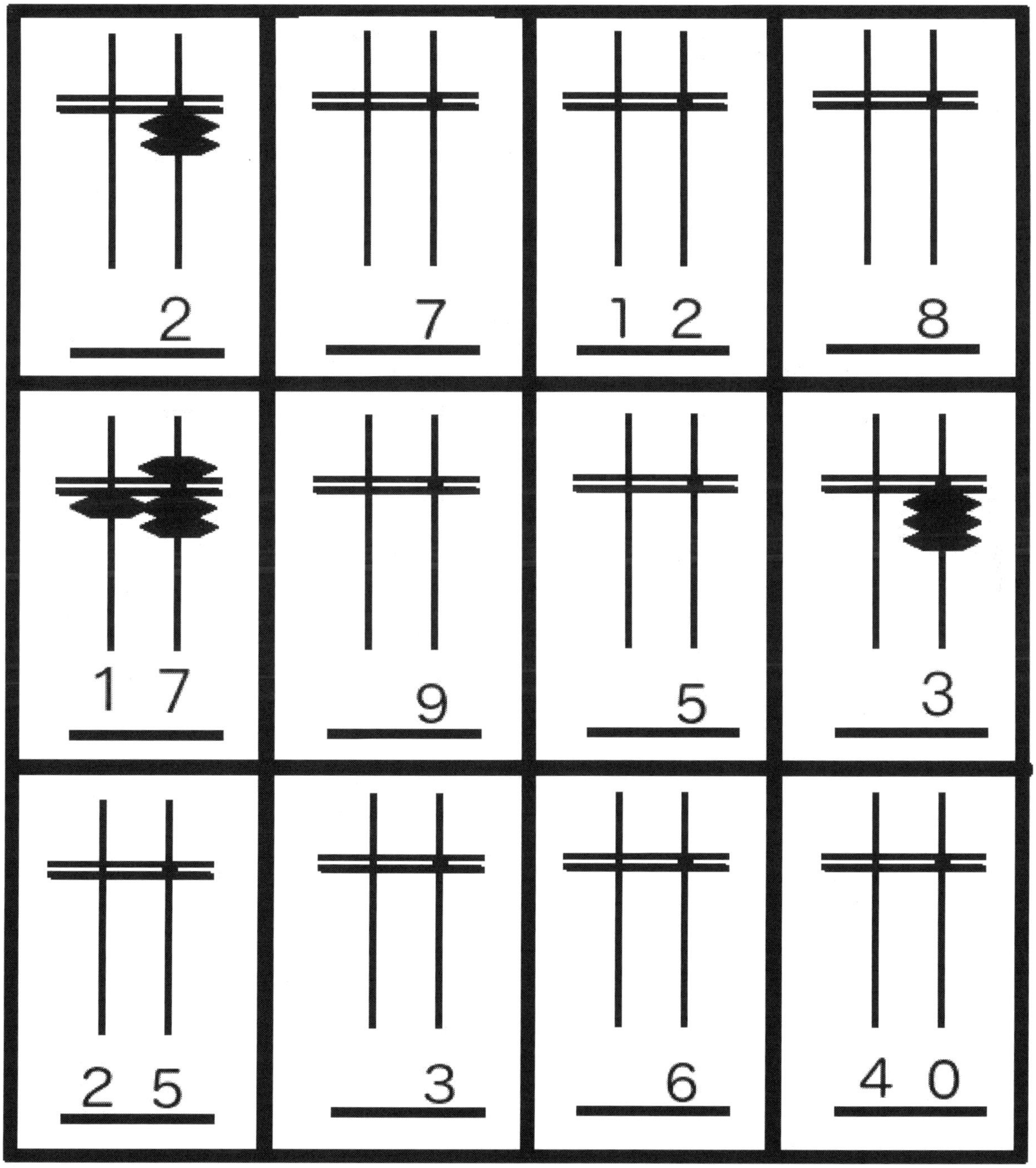

Wow, nice job!

Let's do some more!

Which one is 8 on the Abacus?

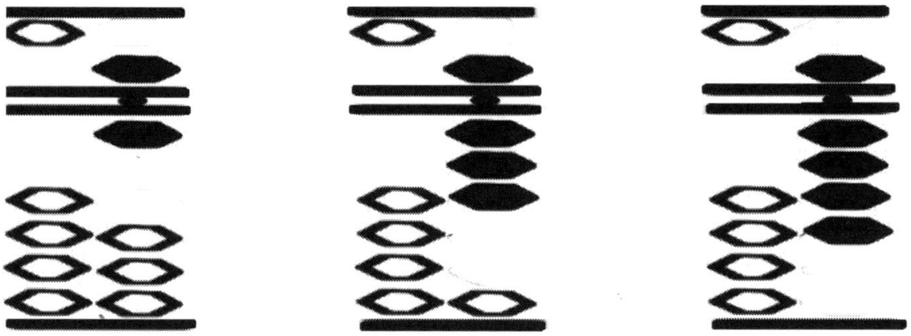

Which one is 4 on the Abacus?

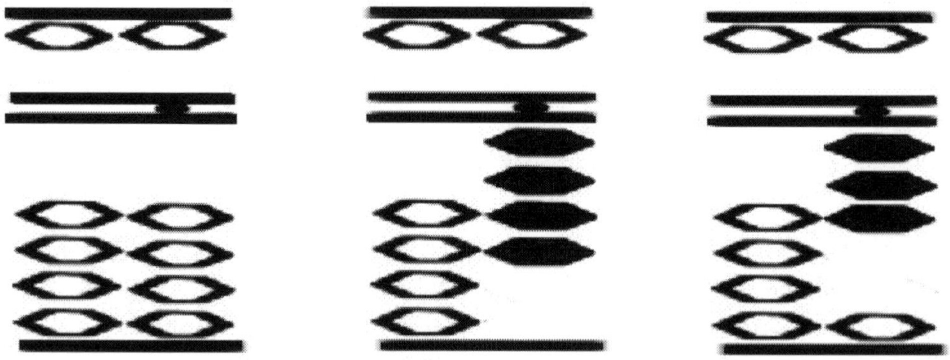

Chapter 3

Abacus Posture

Sit with your back straight while solving questions with the Abacus.

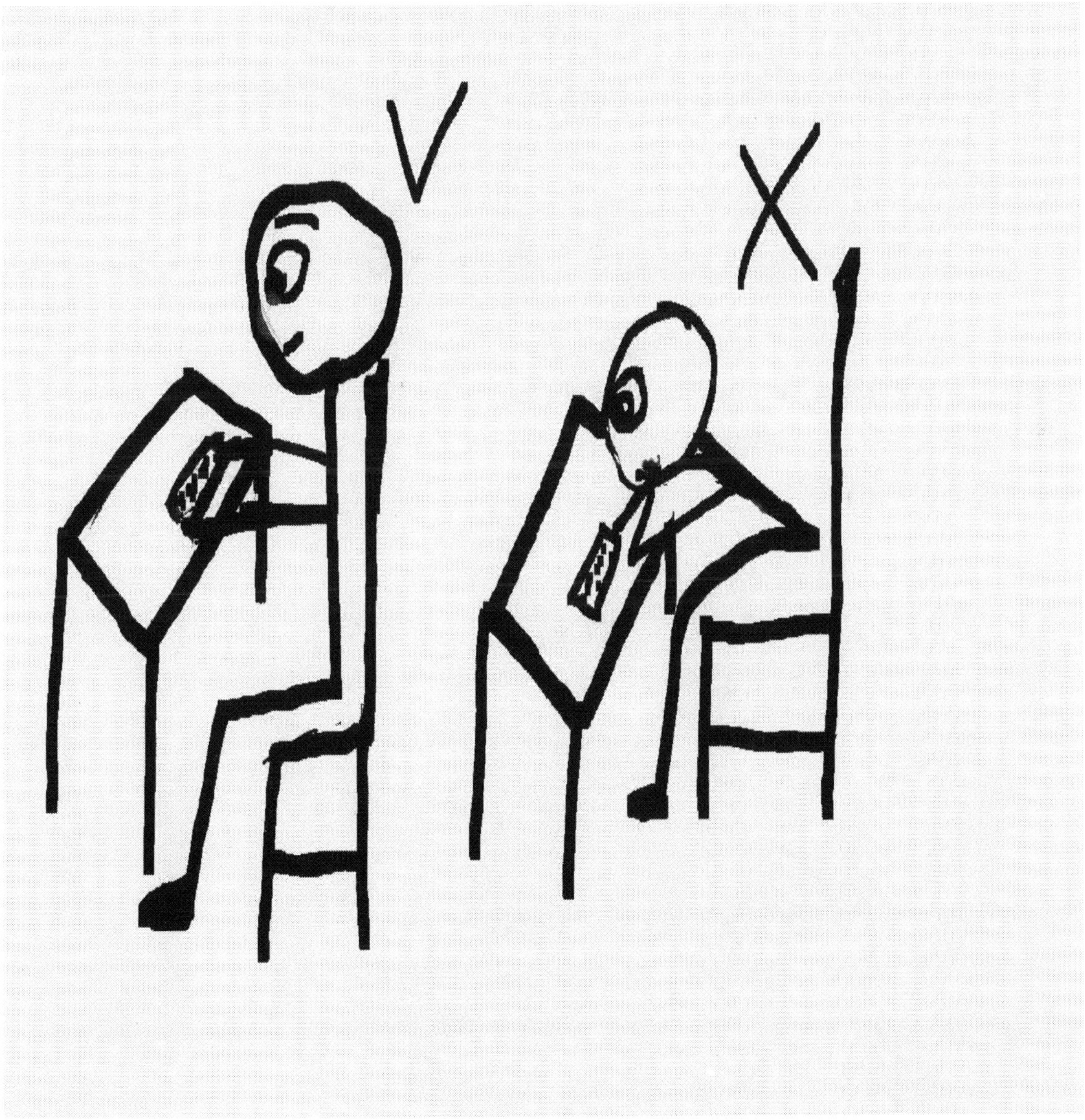

Chapter 4

How to use Abacus

It does not matter if you are right handed or left handed as long as you use one hand to correctly hold the Abacus and the other hand's fingers to move the beads efficiently.

How to hold the Abacus correctly

Use your 5 fingers on your non-dominant hand to firmly hold the Abacus while keeping your elbow off the table. Tilt the Abacus a little bit upward to shift down to read the next number.

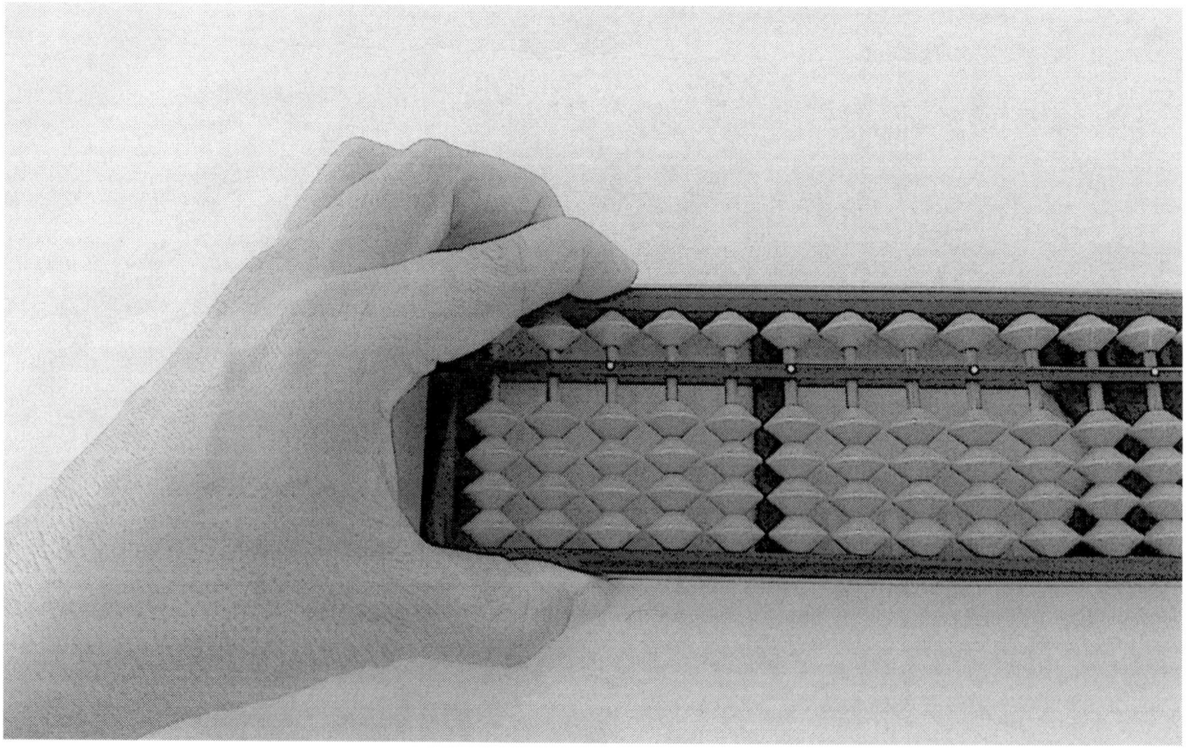

How to move the beads efficiently

Use your thumb and index finger on your dominant hand to move the beads while the other three fingers hold the pencil.

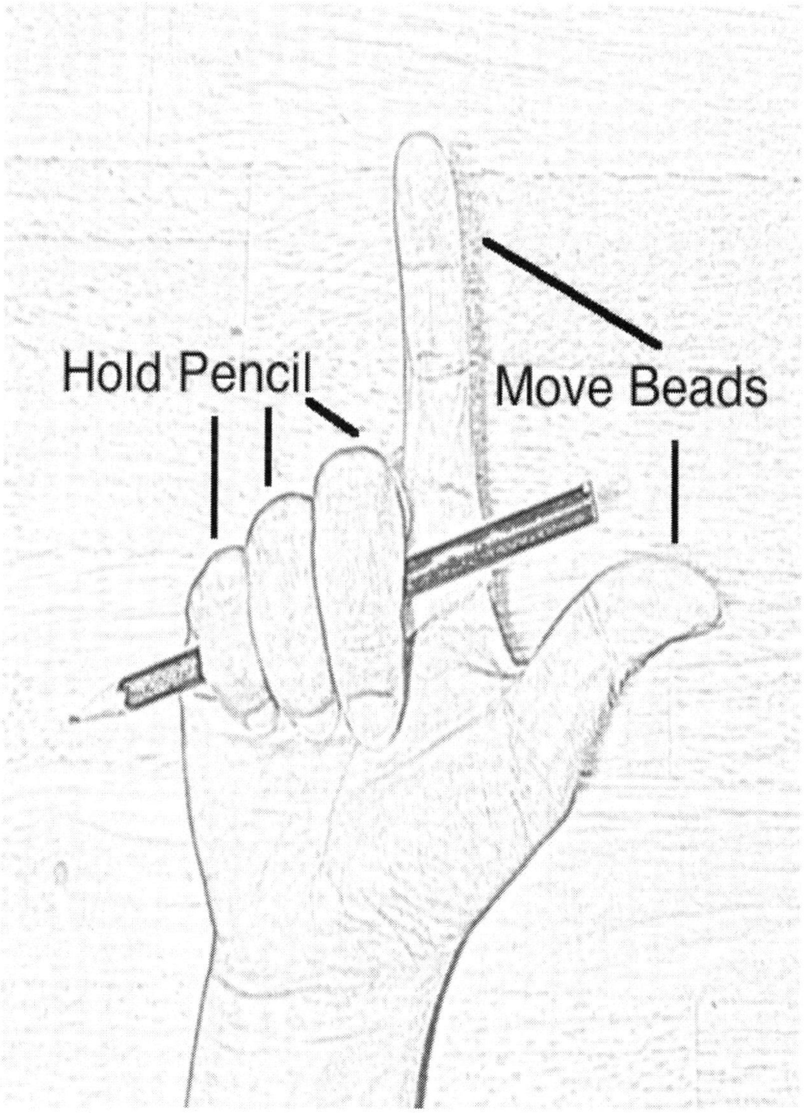

How to clear Abacus

Before calculating, your thumb and index finger need to move from right to left along the beam to clear the Abacus.

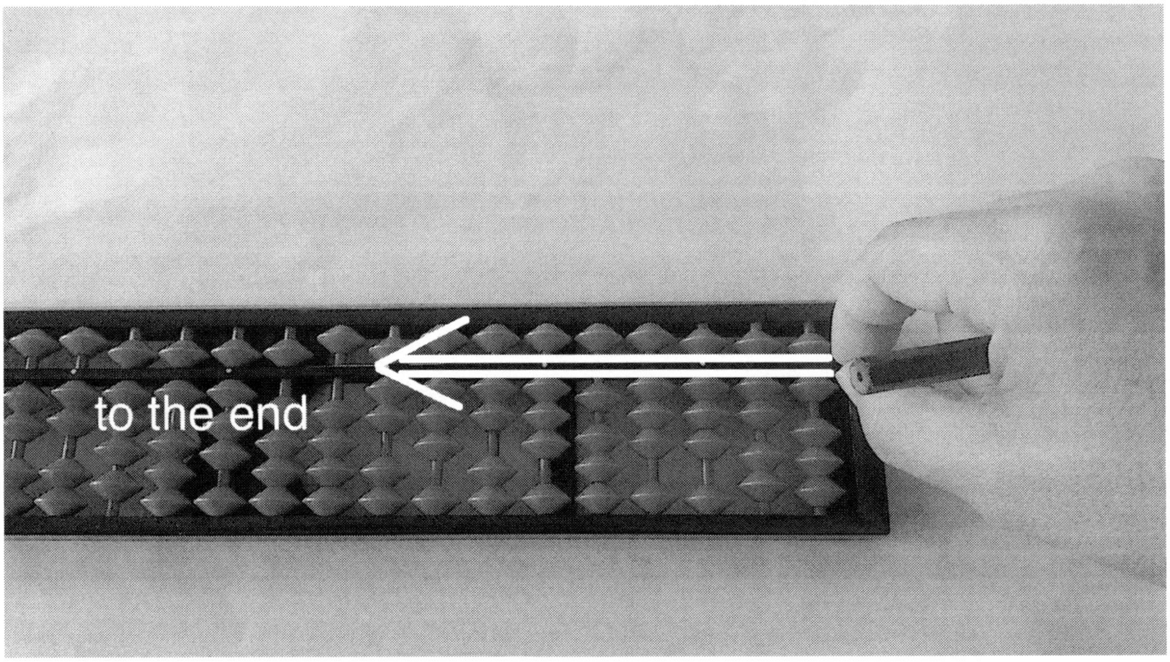

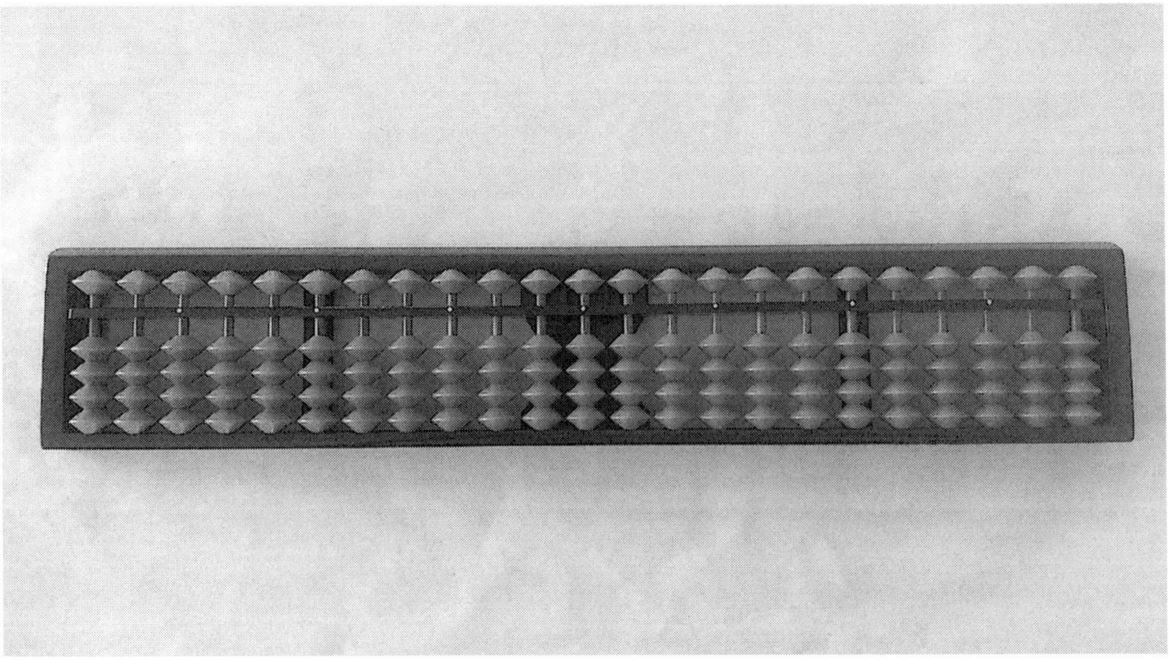

Chapter 5

Finger Rules

Thumb: always moves up beads below the beam. It happens when adding numbers.

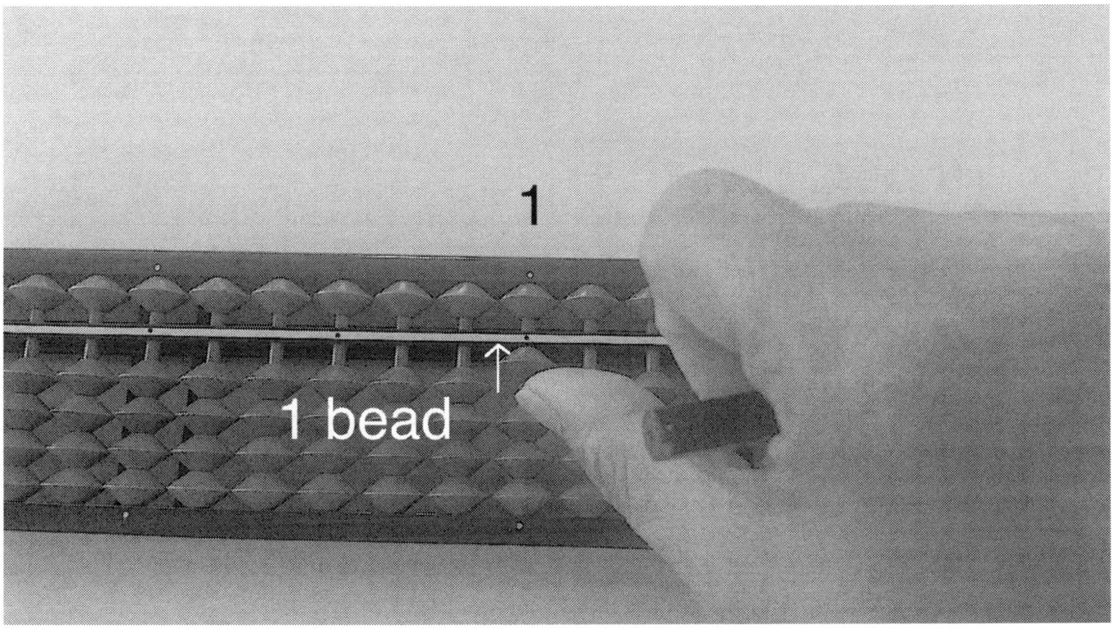

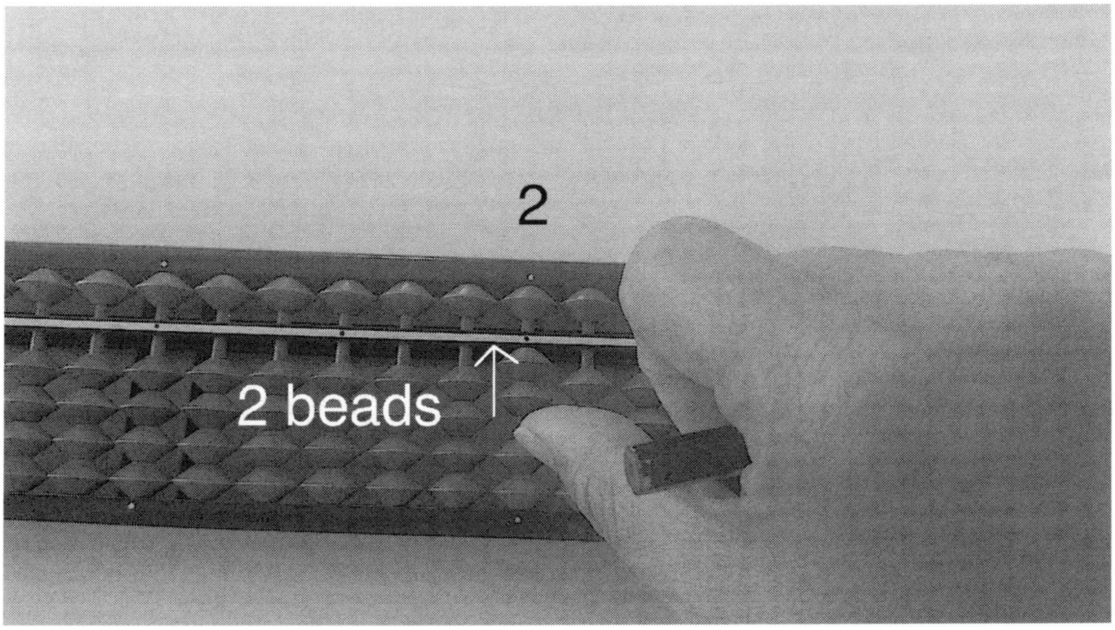

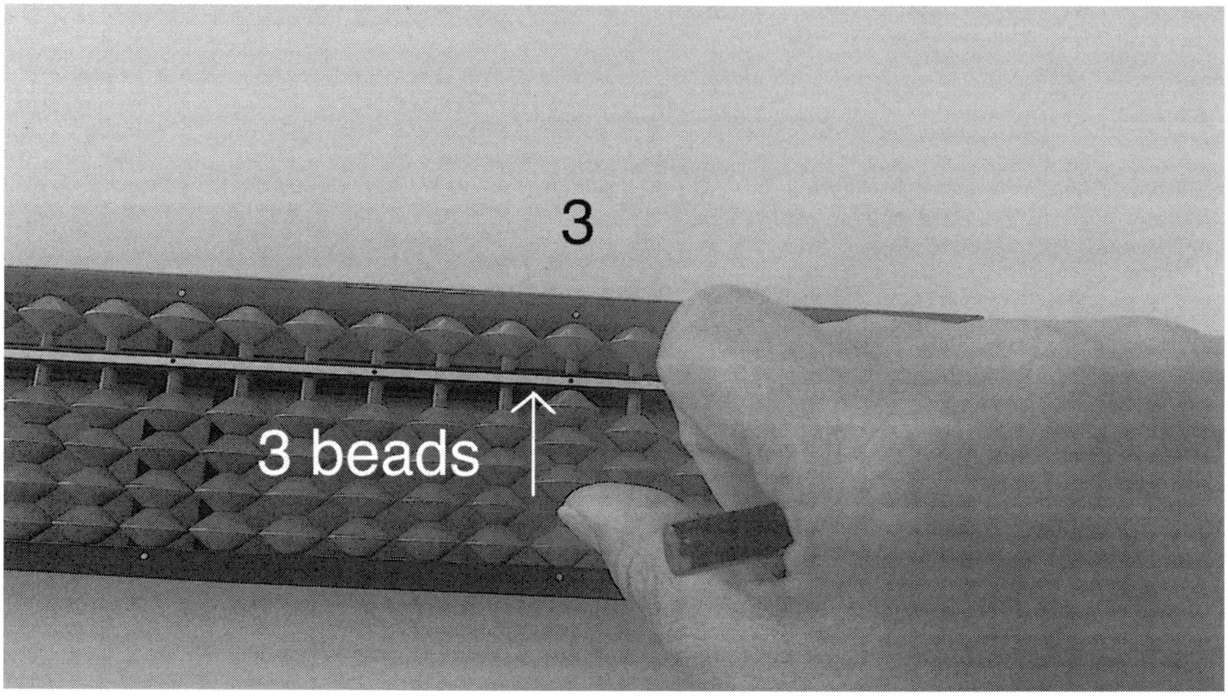
3
3 beads

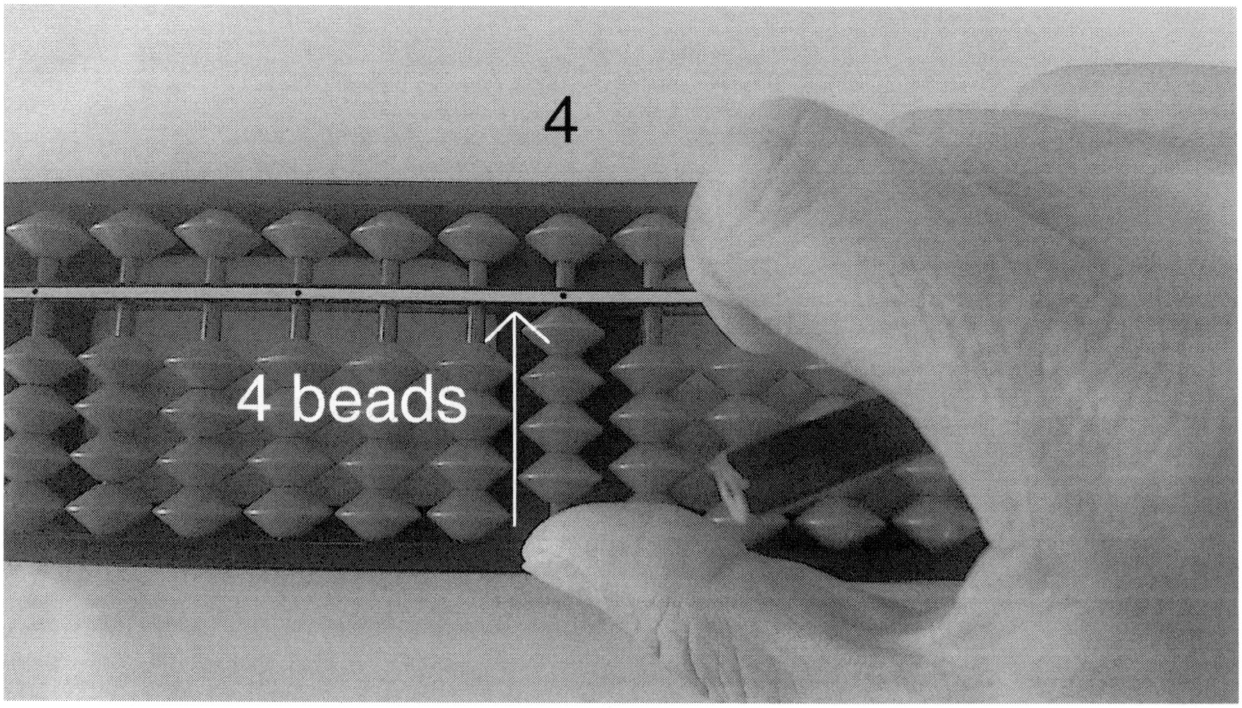
4
4 beads

Index Finger: **A**. moves down the bead above the beam. It happens when adding 5.

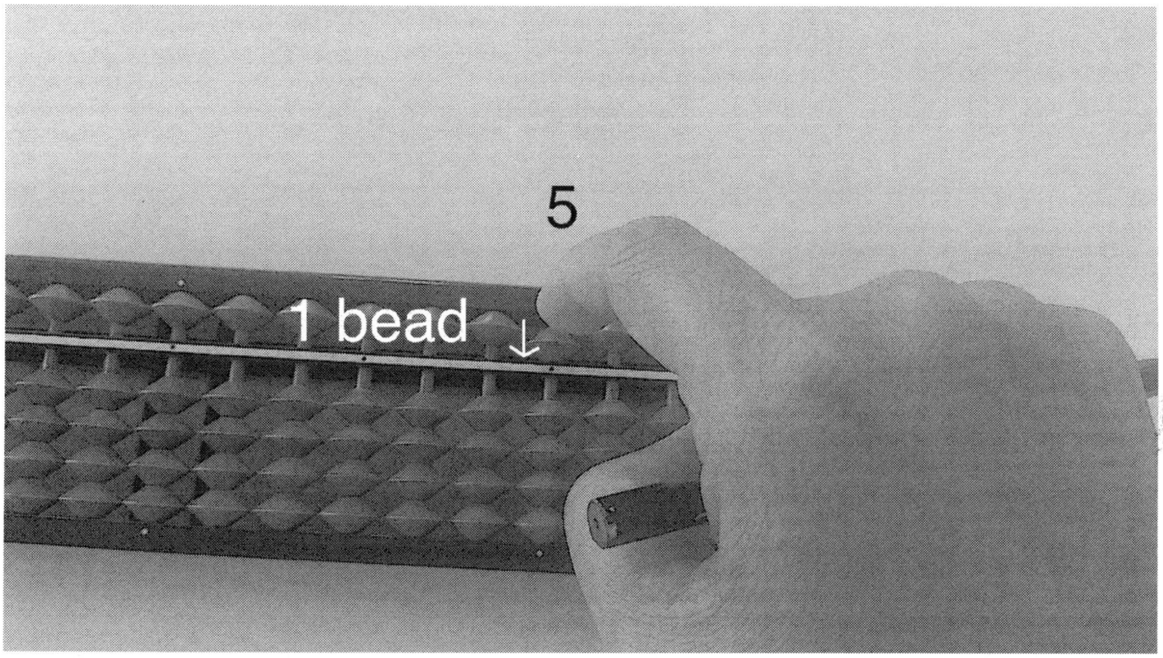

Index Finger: **B**. moves up the bead above the beam. It happens when subtracting 5.

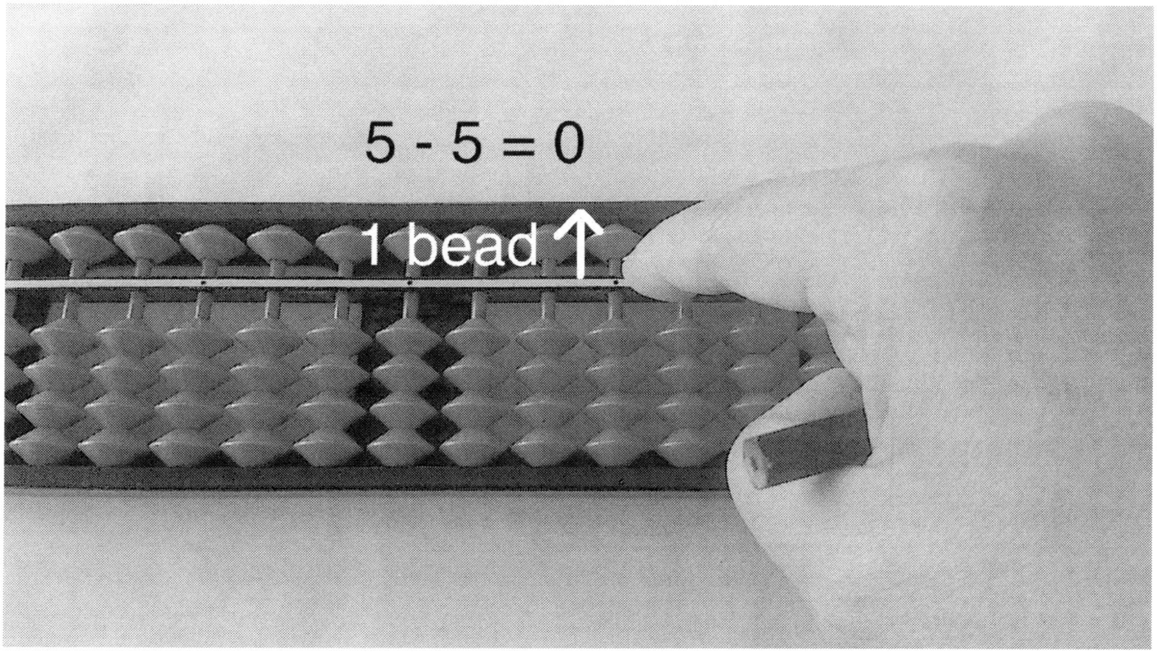

Index Finger: **C**. moves down beads below the beam. It happens when subtracting numbers.

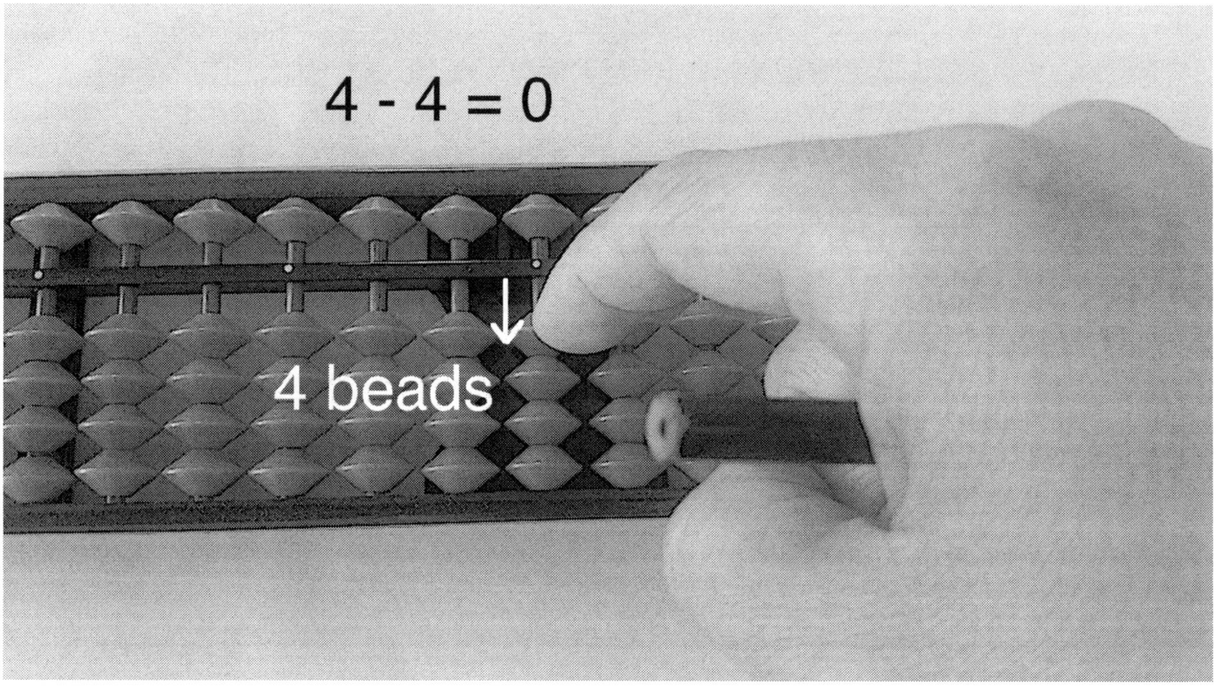

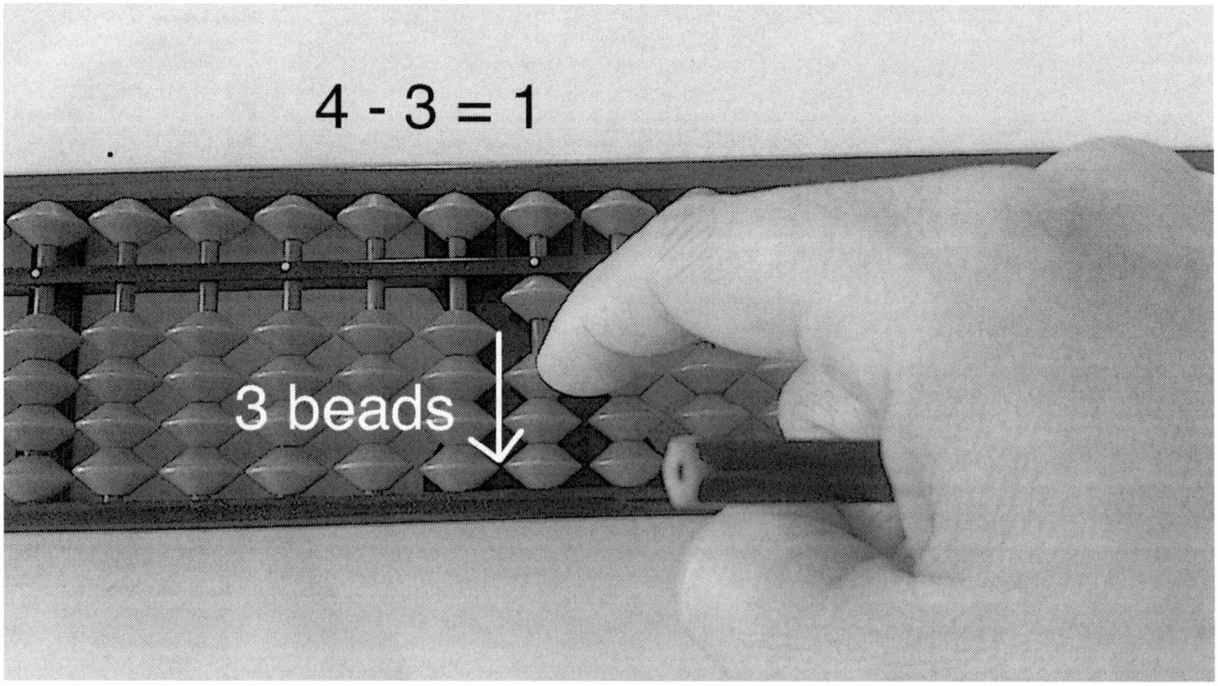

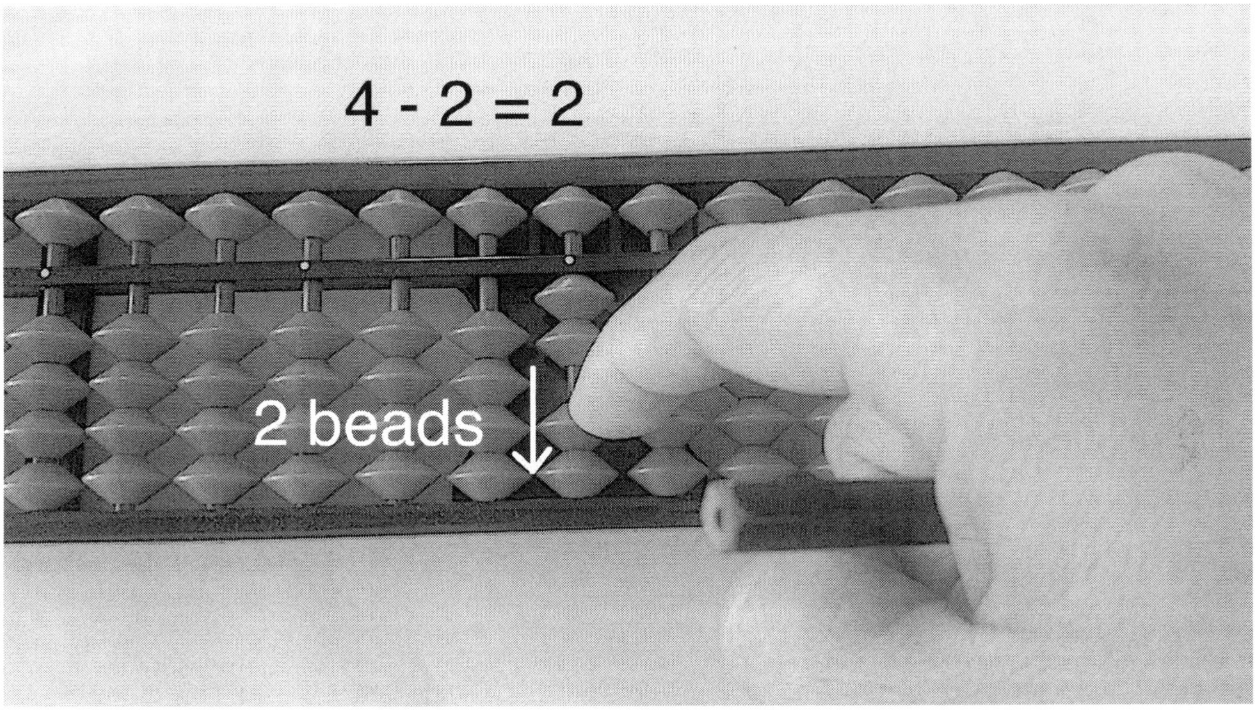
4 - 2 = 2
2 beads

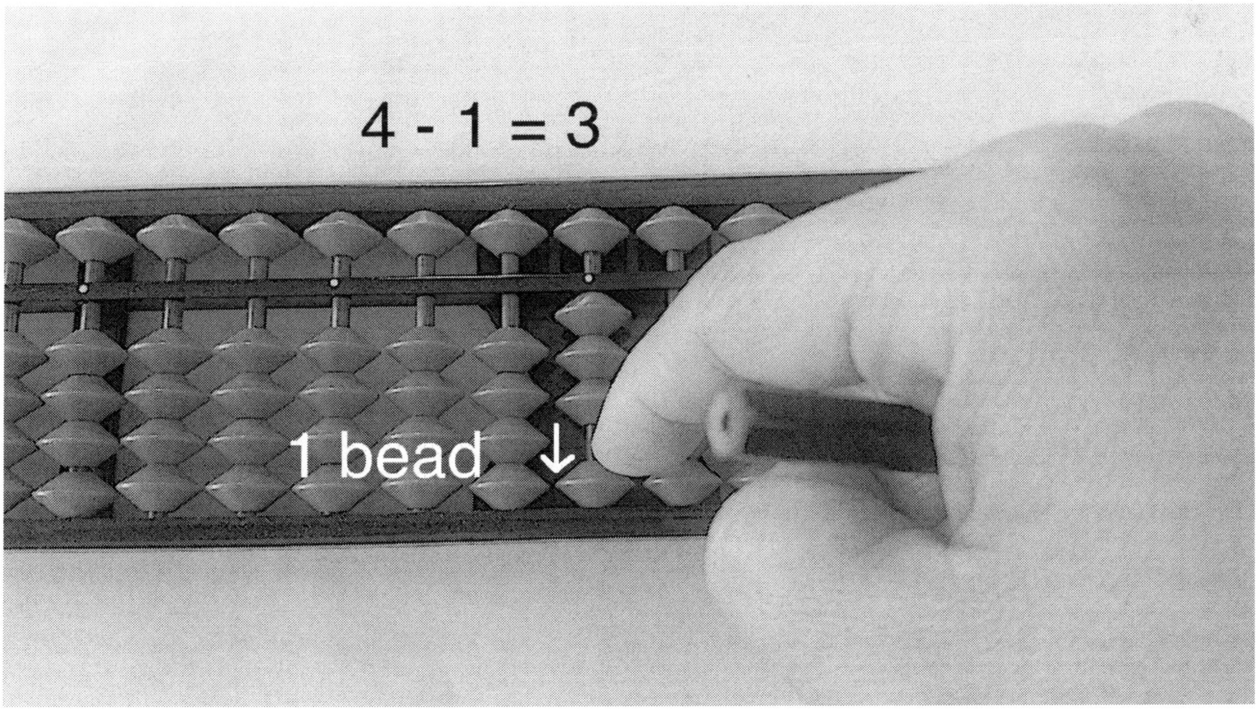
4 - 1 = 3
1 bead

Thumb & Index Finger:

A. Move beads inward at the same time.

When thumb and index finger are moving beads inward at the same time, large numbers such as 6, 7, 8 and 9 will show up. It happens when adding numbers.

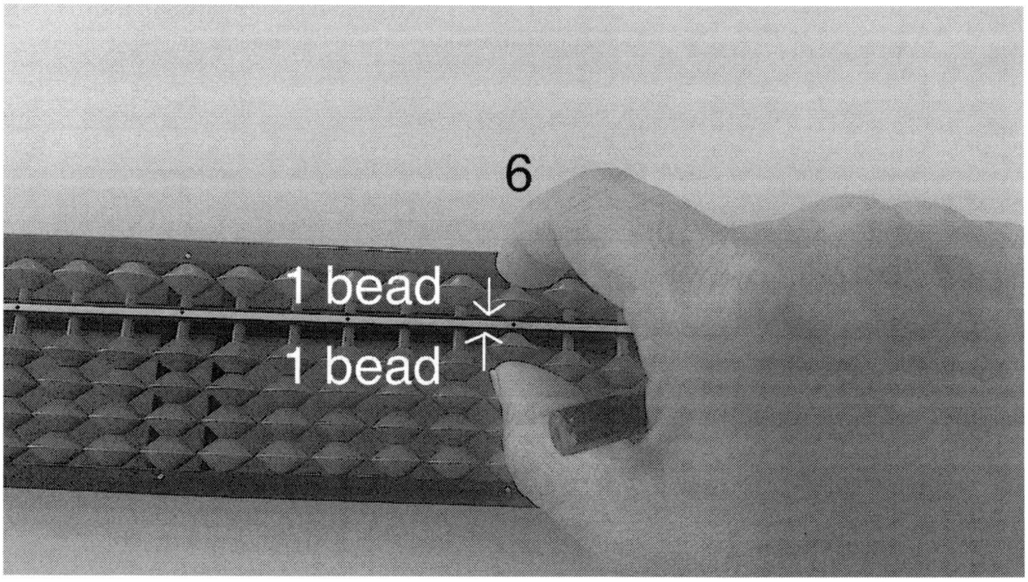

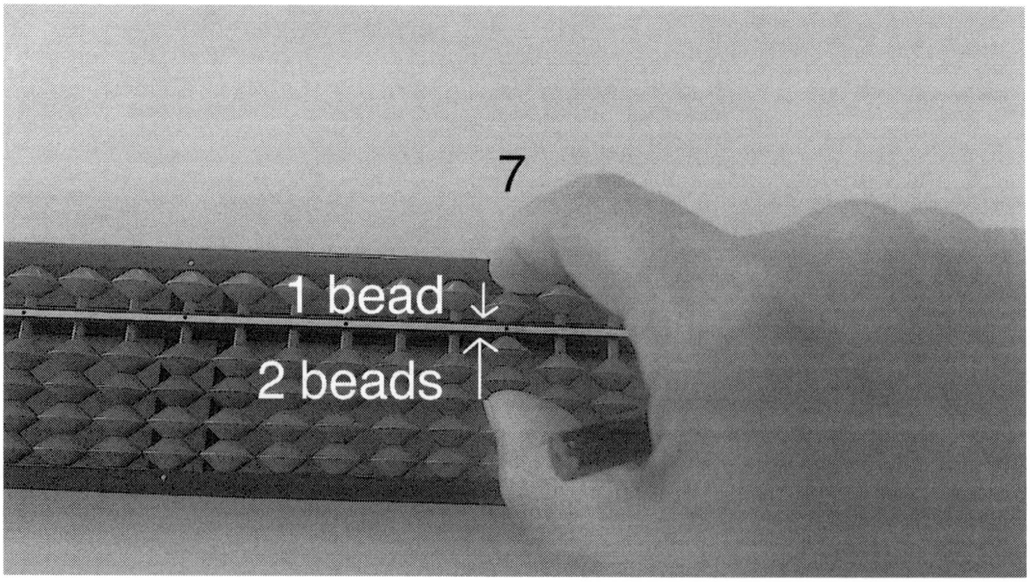

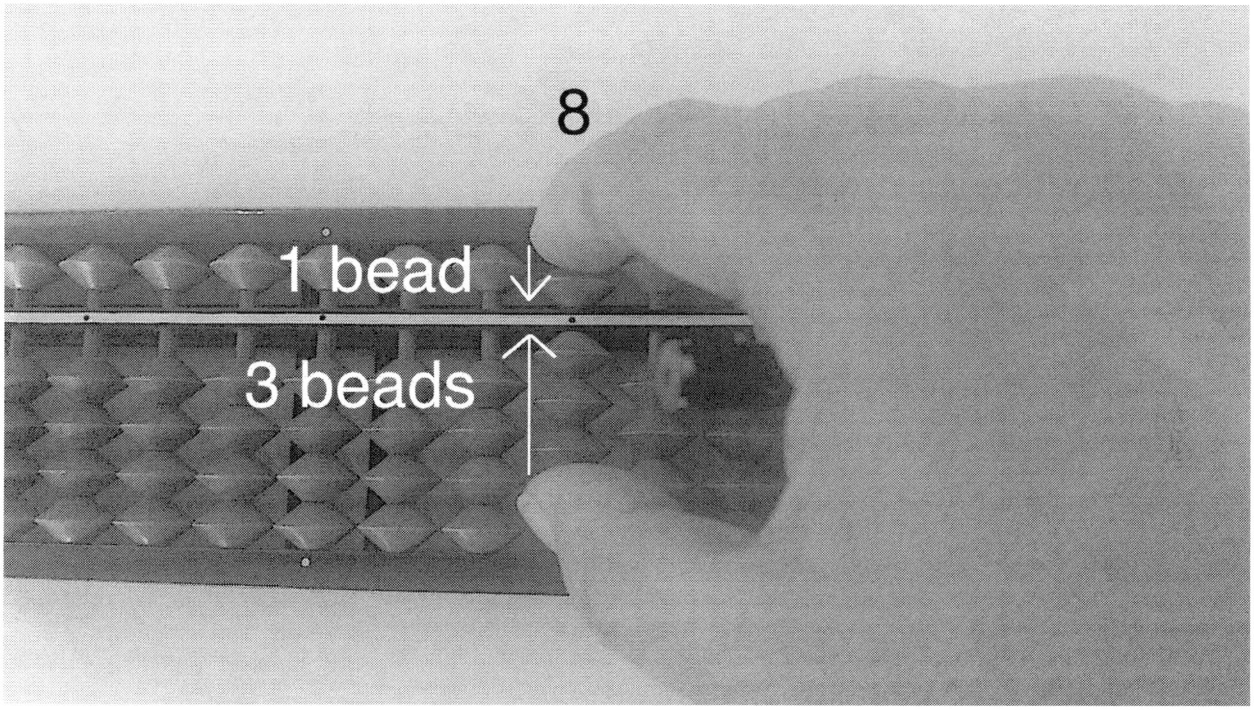
8
1 bead
3 beads

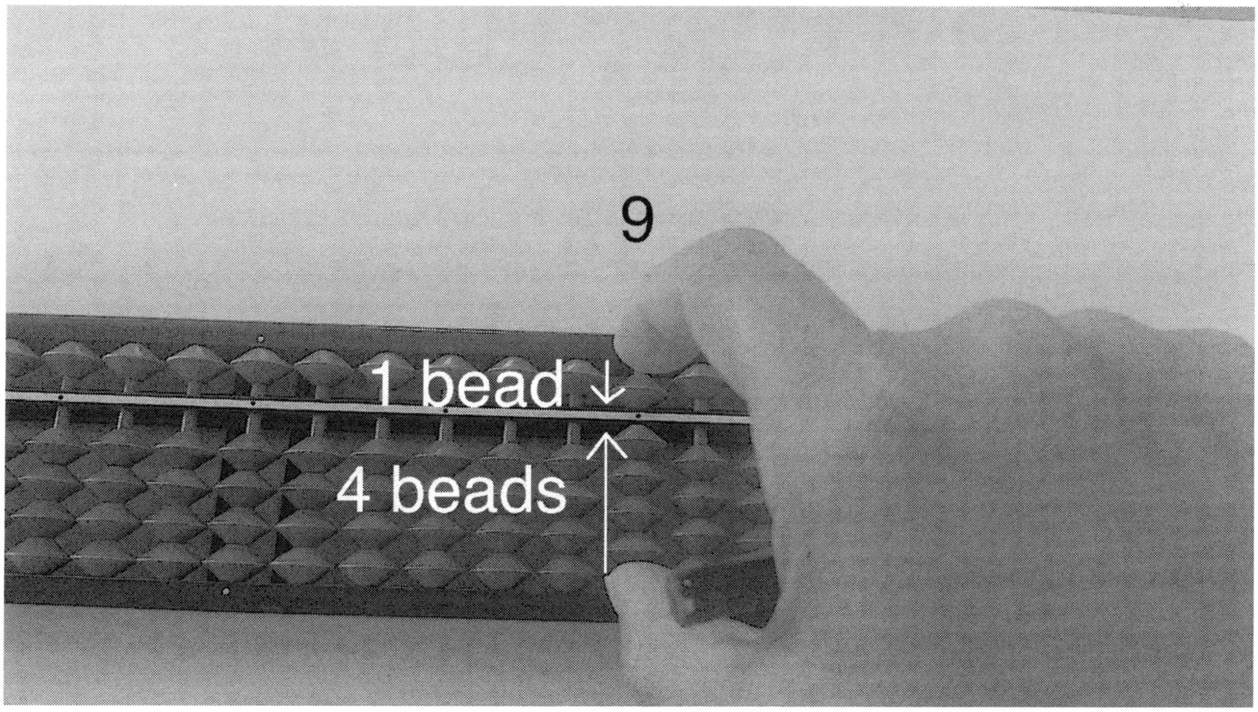
9
1 bead
4 beads

Thumb & Index Finger:

B. Move beads outward at the same time.

When thumb and index fingers are moving beads outward at the same time, it means to subtract numbers.

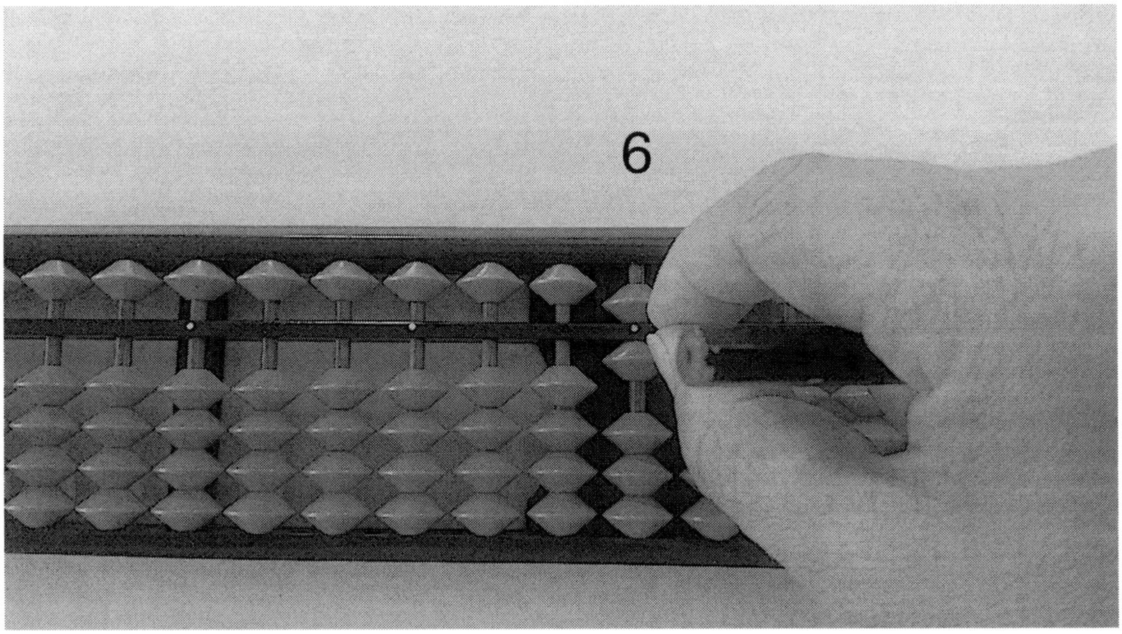

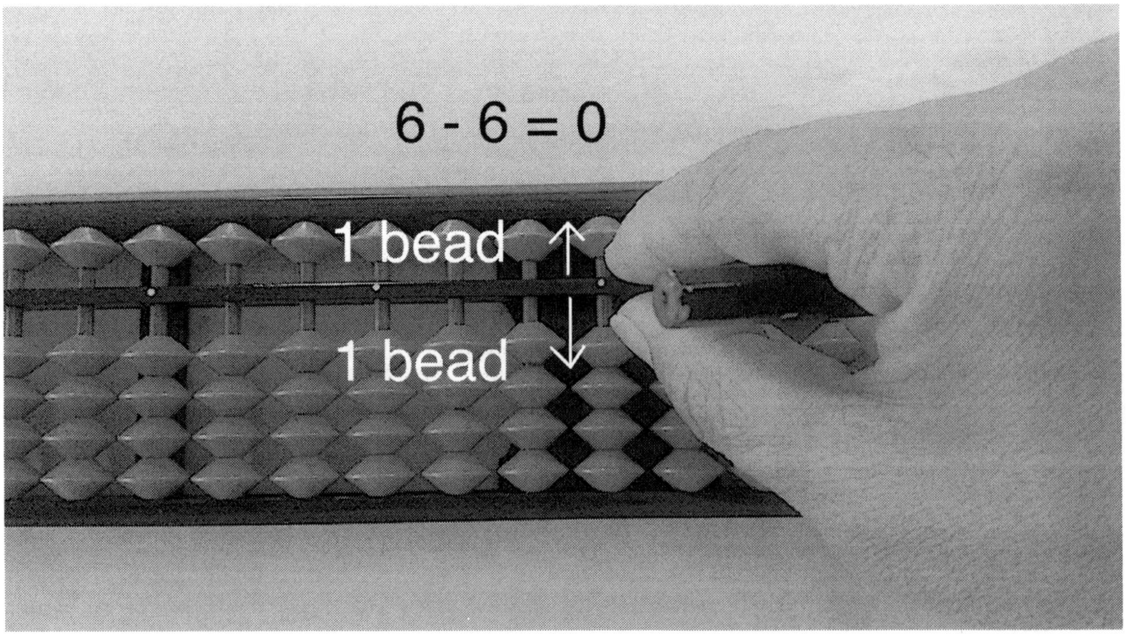

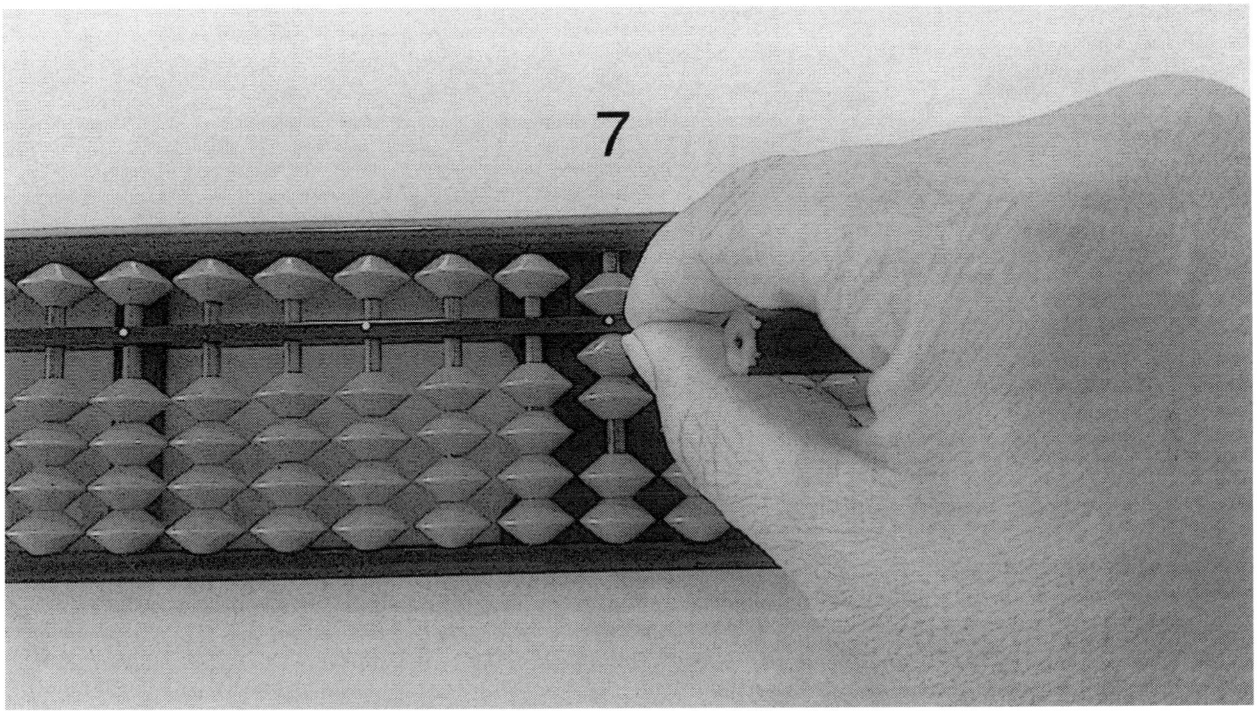
7

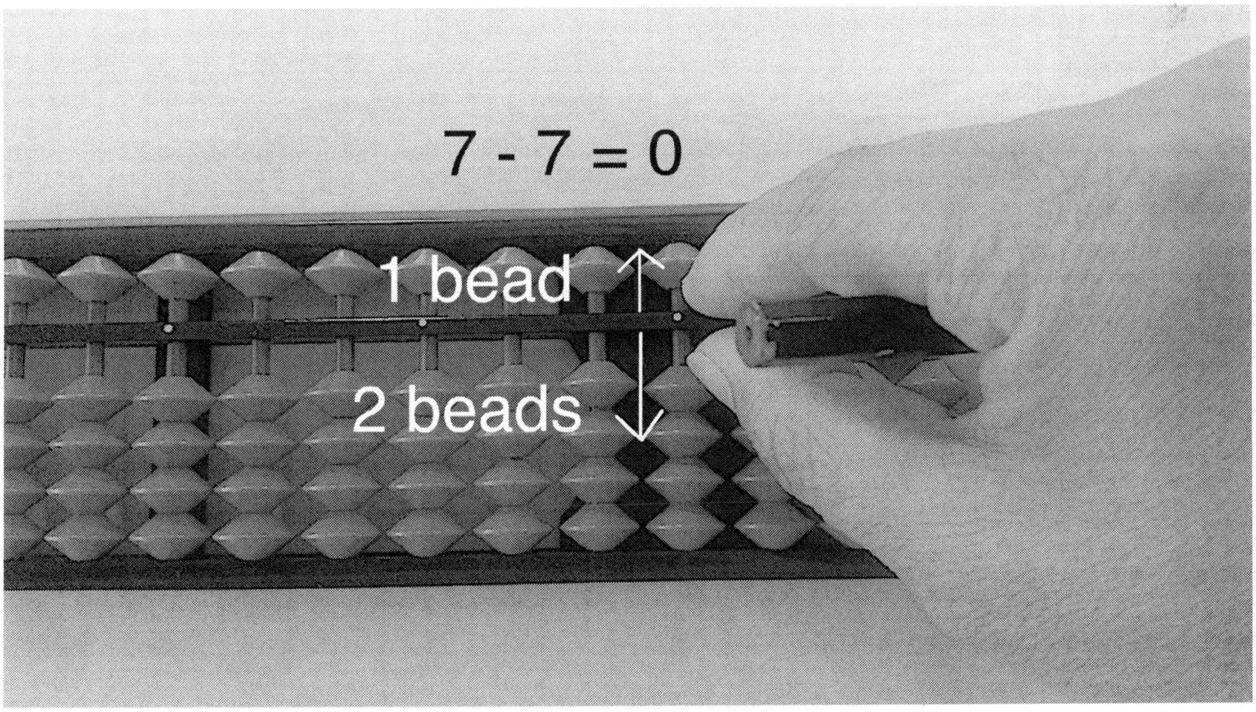
7 - 7 = 0
1 bead
2 beads

Thumb & Index Finger move outward

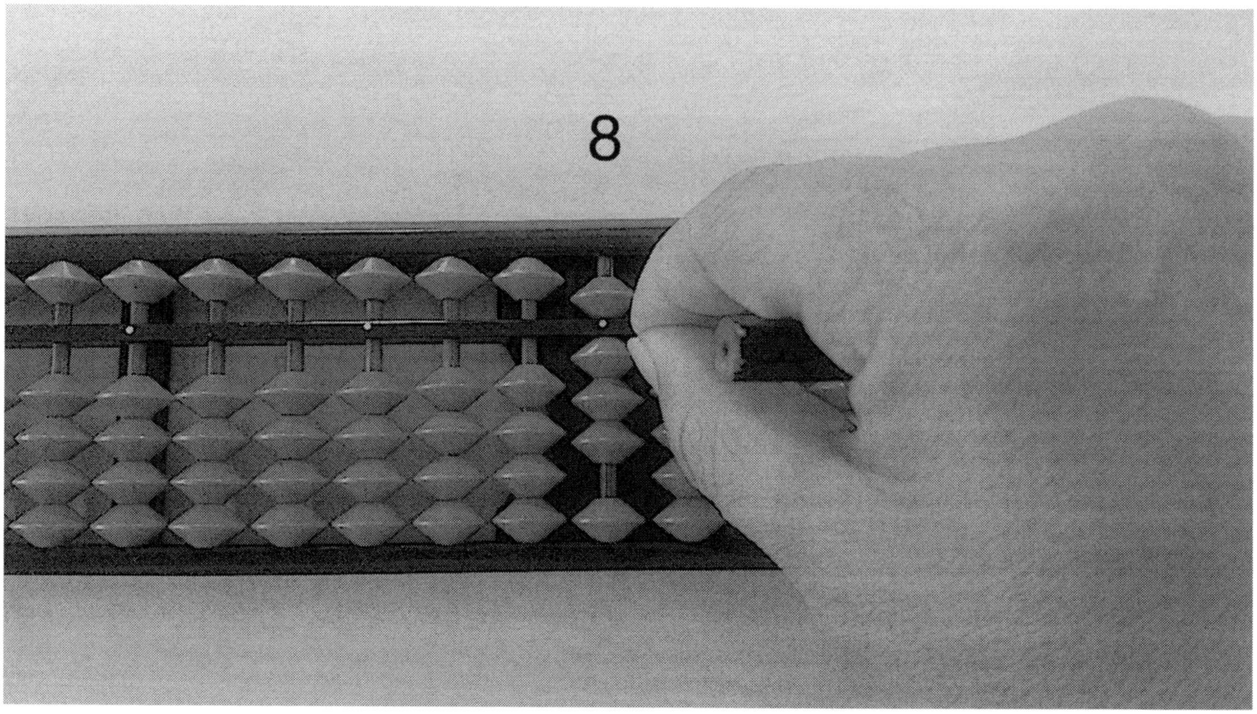

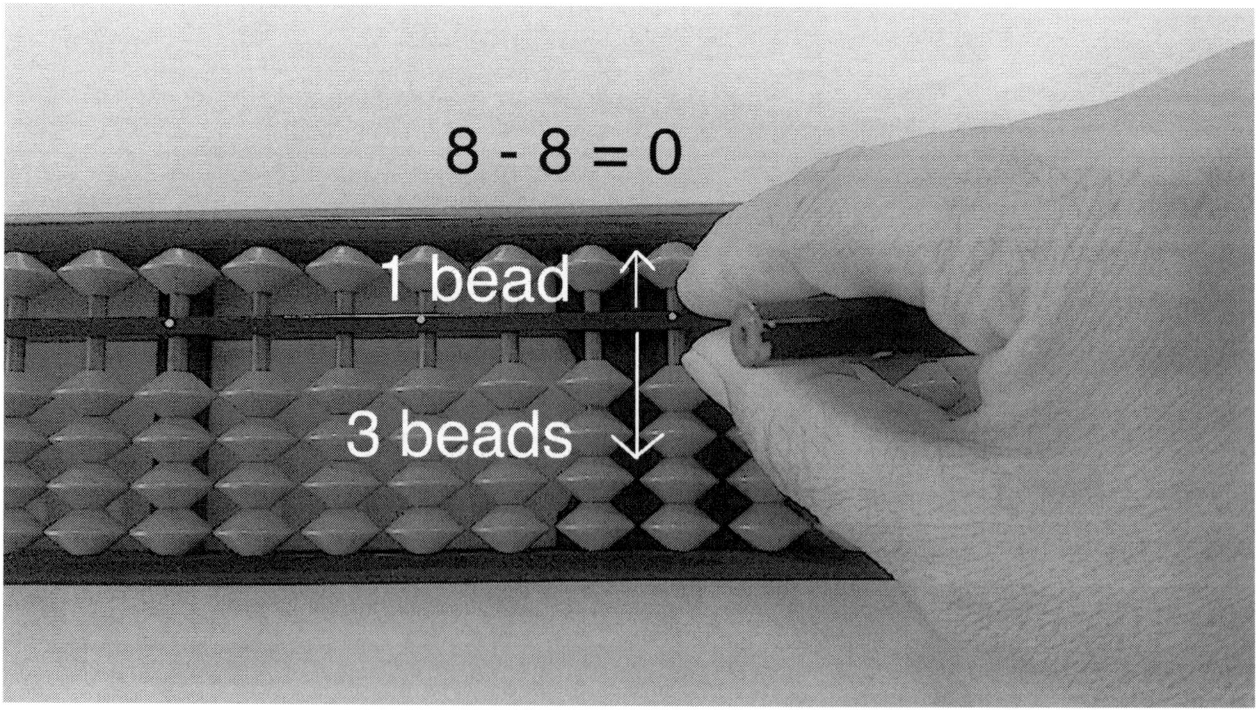

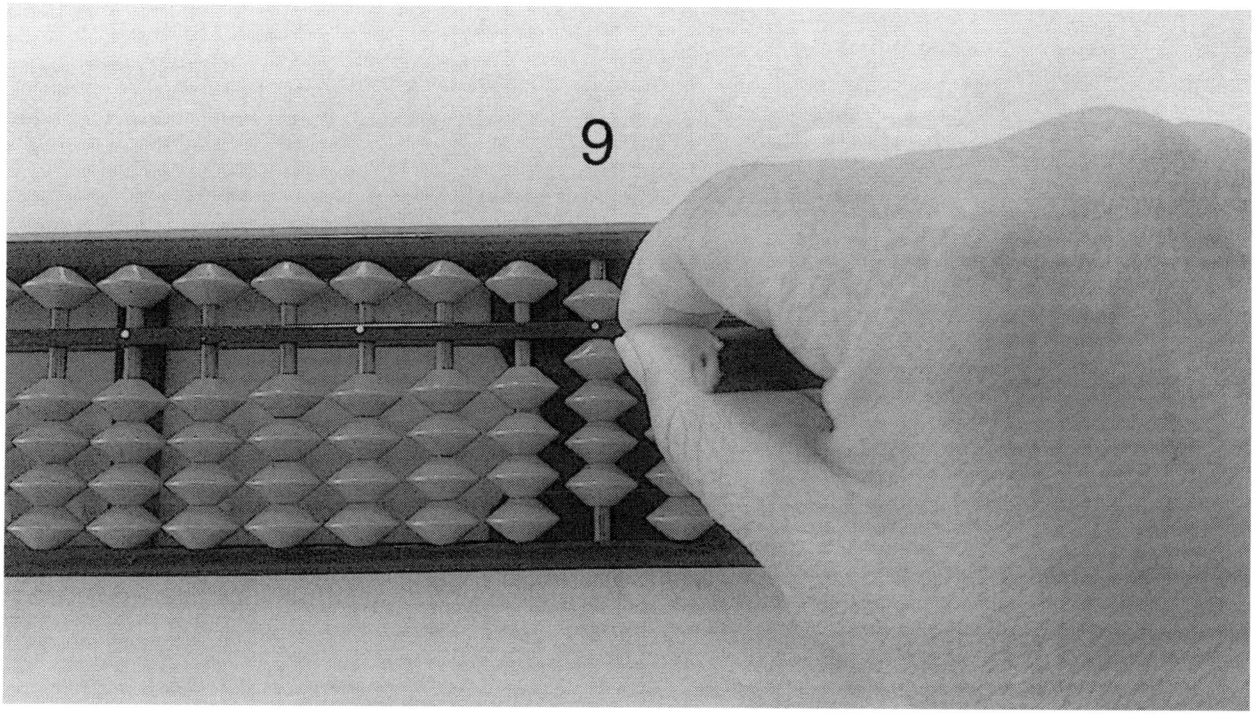
9

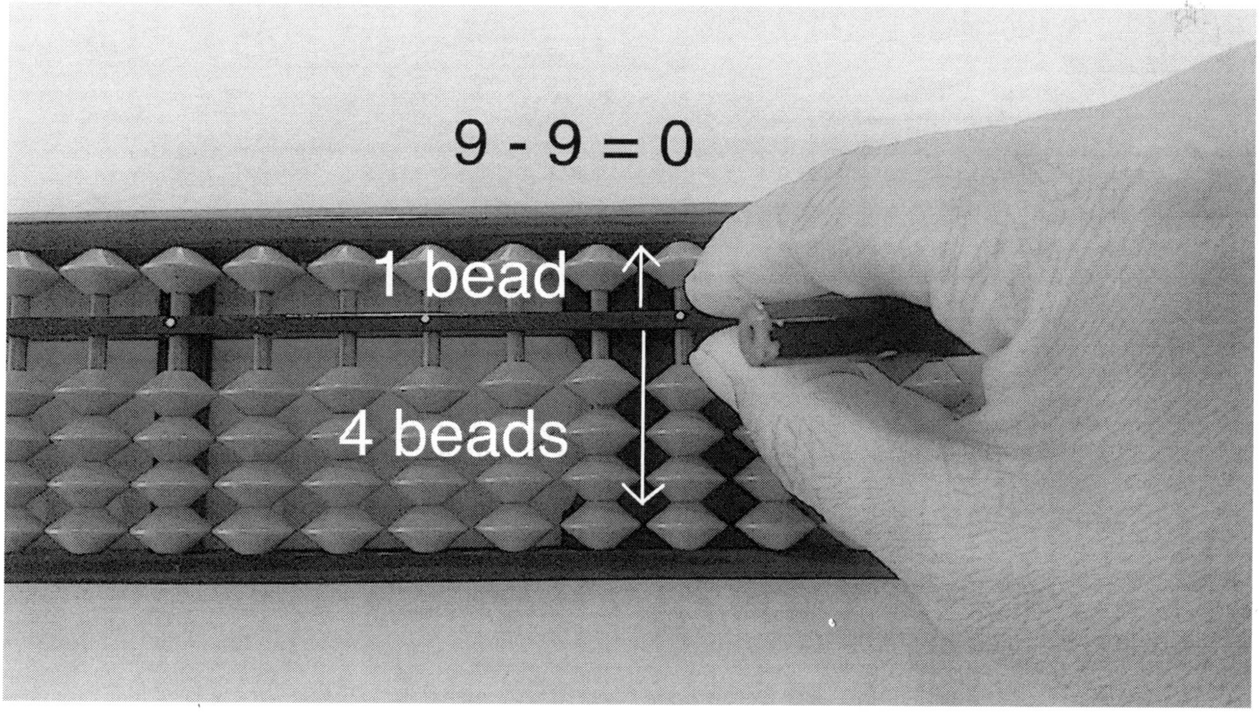
9 - 9 = 0
1 bead
4 beads

One Movement for upper & bottom beads

One Movement is to move the upper and bottom beads at the same time to get answers quickly. It involves with formulas. Please refer to “How to Abacus 2: All About Formulas”.

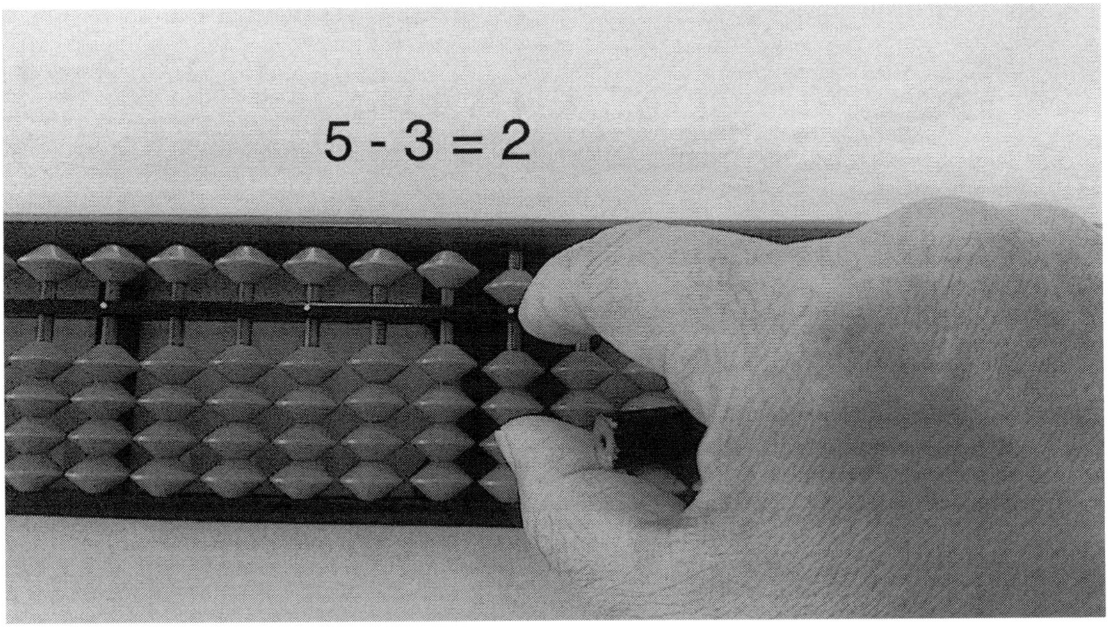

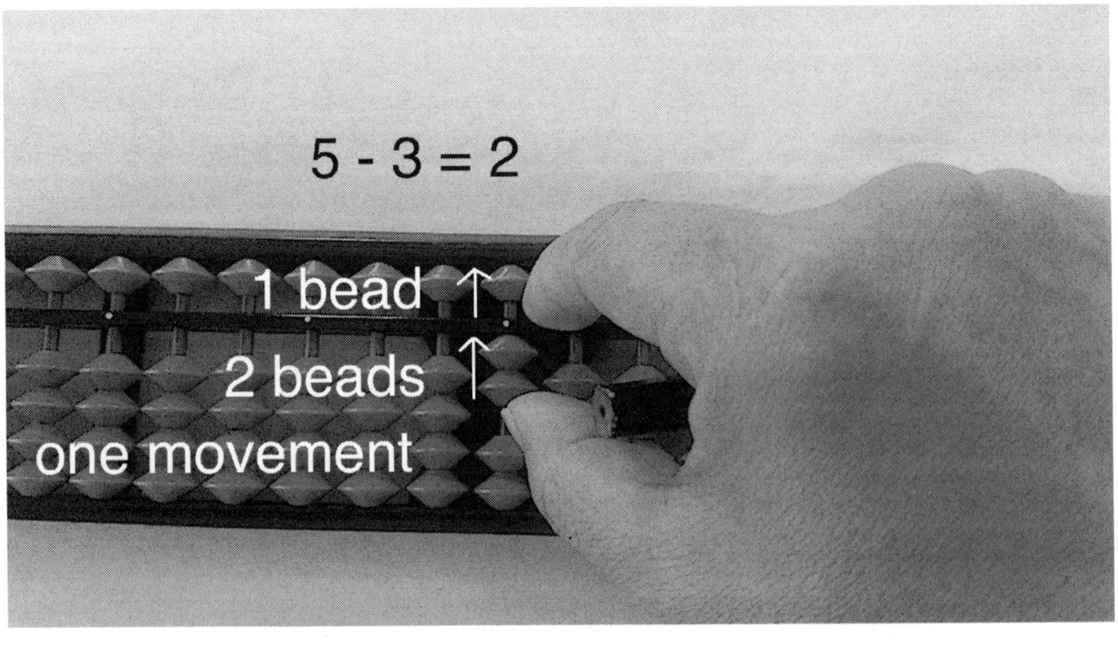

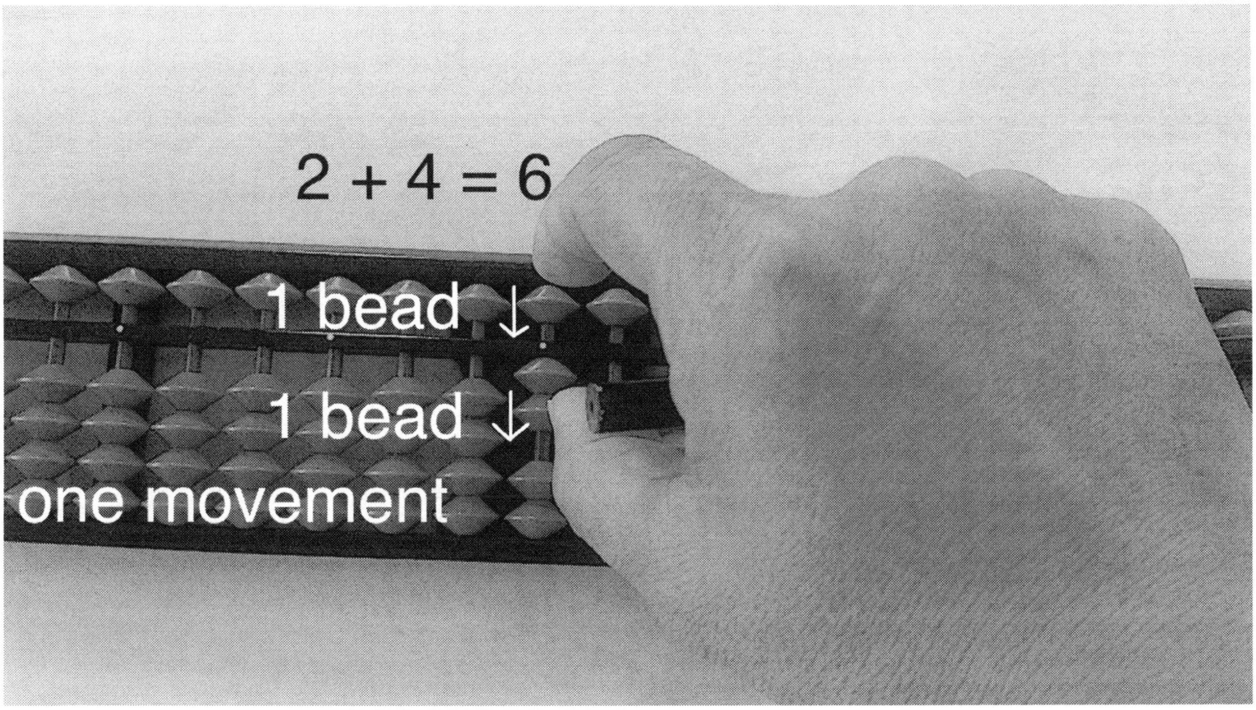
2 + 4 = 6
1 bead ↓
1 bead ↓
one movement

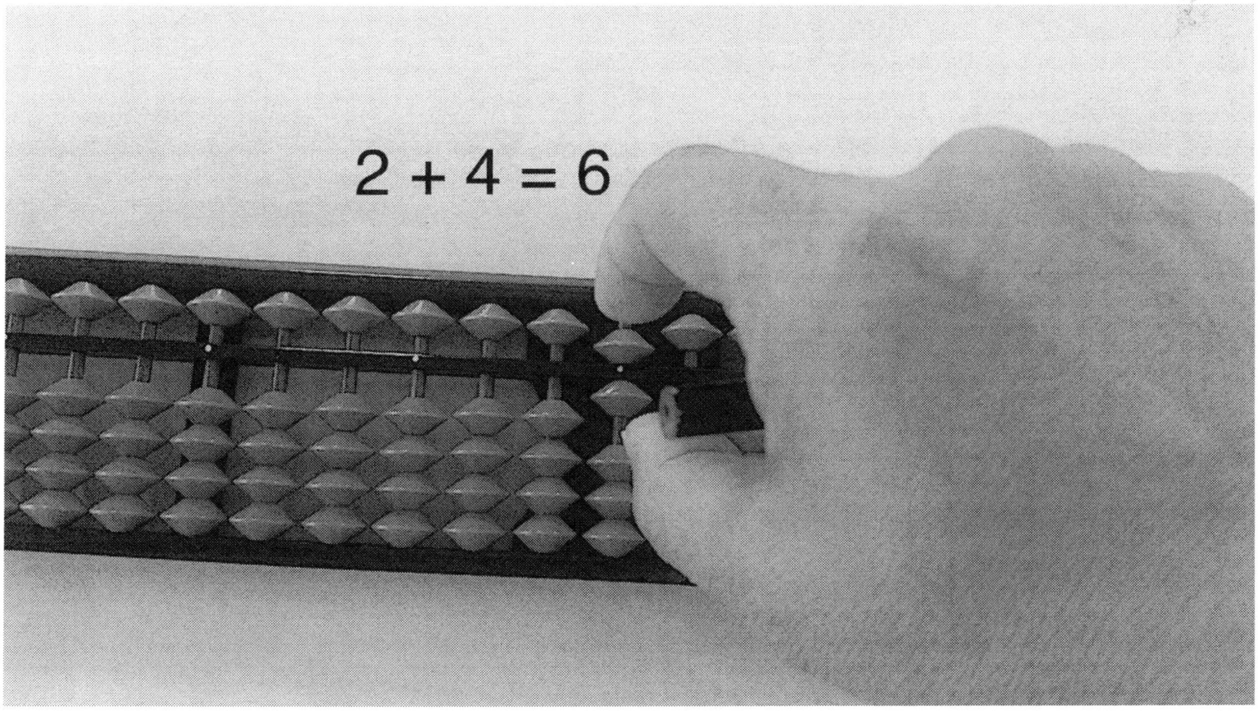
2 + 4 = 6

The order of moving beads

Just like the way we write numbers, always move beads from left to right. For instance, in order to show 12 on Abacus first move 1 on tens place, then move 2 on ones place.

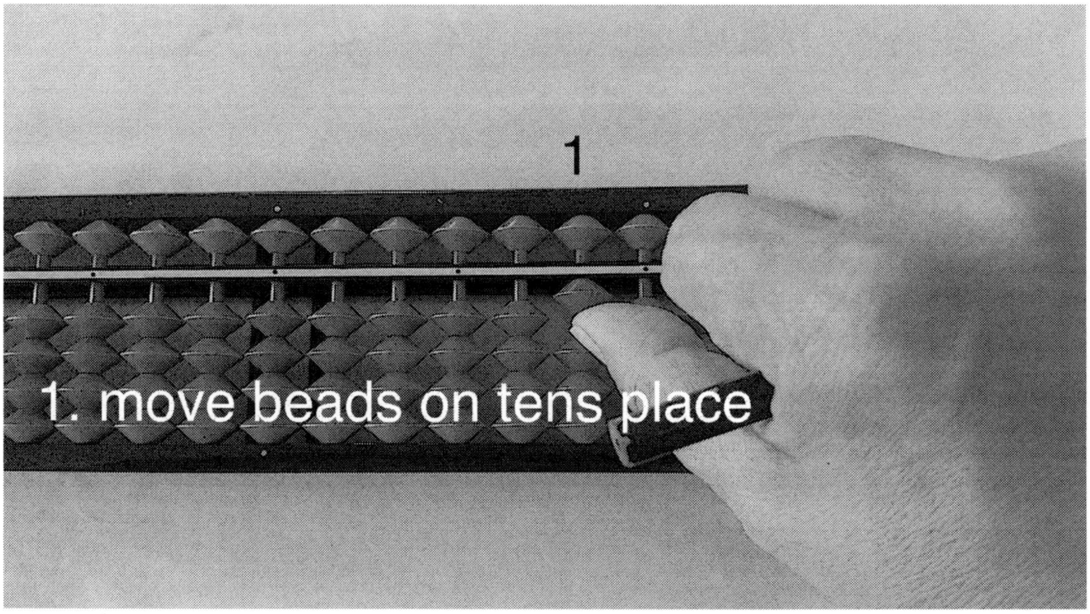

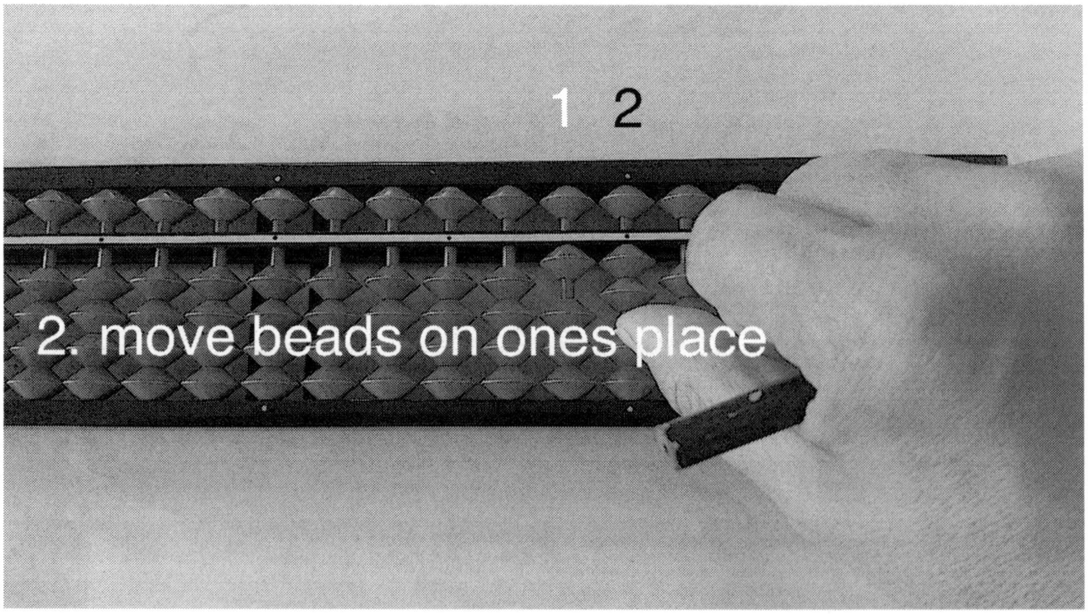

Time to exercise Fingers!

Which finger(s) should you use for these numbers?

3 7 4 5 9 2 8

Shift down Abacus to read the next number

This usually happens when solving addition and subtraction problems. Shifting Abacus down when reading the next number will help you focus on the number you are calculating. The following picture is an example of Abacus problems involving one digit, 10 numbers.

1	2	3	4	5	6	7	8	9	10
5	9	8	1	8	4	6	2	7	3
8	6	3	5	9	7	2	1	4	3
-2	1	5	-4	-6	4	-1	7	8	5
9	-5	6	9	6	8	3	5	1	-2
7	6	-3	8	-5	8	-6	-9	5	8
5	3	4	8	9	3	5	3	1	3
-4	6	-2	7	-3	4	-1	7	3	6
5	8	7	-9	5	-5	9	-3	6	7
3	-5	7	-3	8	6	8	4	1	8
6	2	-4	6	8	6	9	5	2	3

Here is another example of Abacus questions with more digits, as shown in the picture. It might be harder to locate the number you're calculating if the Abacus is not shifting down to the next number.

1	2	3	4	5
2,345.76	543,445.54	32,650.88	5,887.77	654,765.90
13,651.76	6,980.44	7,854.91	563,786.99	46,721.64
876,644.21	534,752.85	652,749.90	98,564.33	3,486.23
6,841.98	87,654.21	-8,760.09	-8,541.18	9,765.17
72,665.53	-9,742.99	265,711.76	760,421.70	765,123.90
543,087.54	651,754.87	86,421.98	7,541.90	76,456.56
8,561.90	-9,876.32	5,921.63	75,765.42	654,989.60
564,721.88	87,323.98	653,839.65	-220,976.56	7,981.50
7,443.65	-541,543.96	9.654.90	110,987.45	654-197.49
654,132.65	8,521.64	276,654.19	-1,989.98	7,543.67

The location of the questions on a page decides where you put the Abacus above or below the problem to focus on the number you are solving. Also, the dot on ones place can help you locate the question you are working on.

First, align the ones place to the question you are solving. Then tilt the Abacus a little bit to shift down to read the next number.

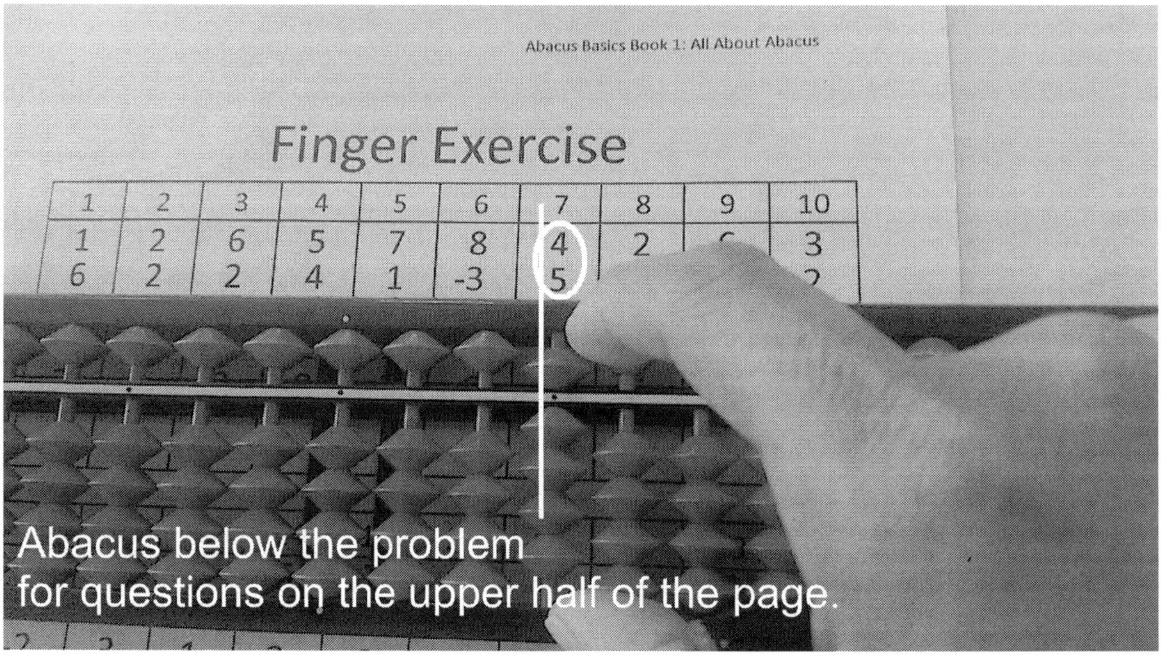

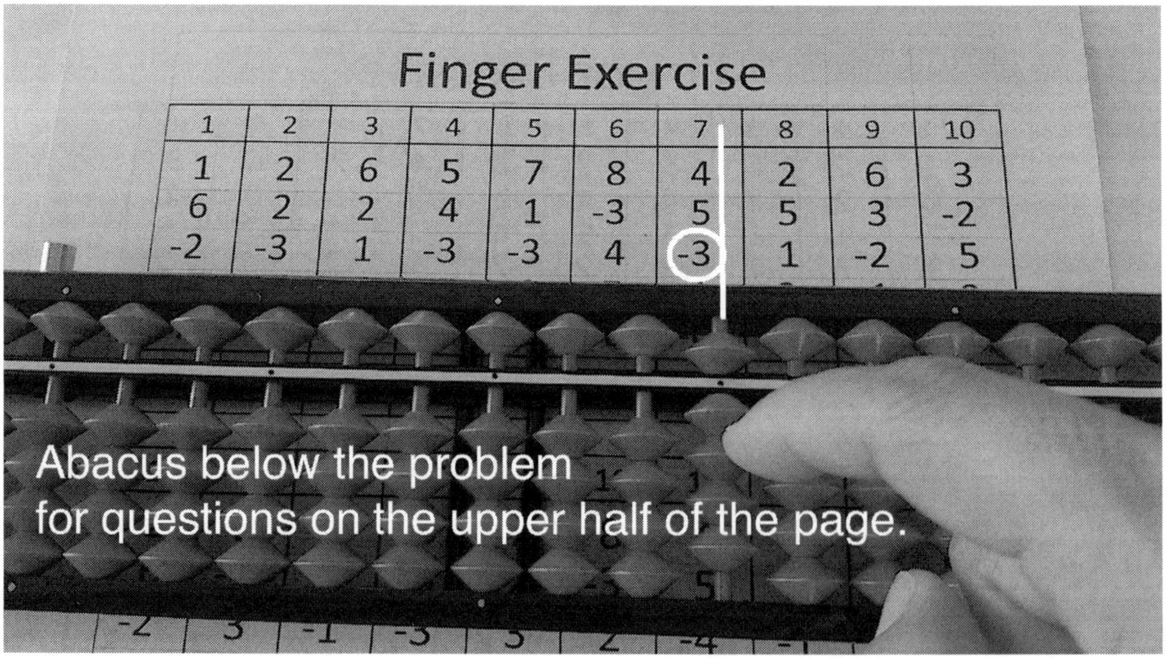

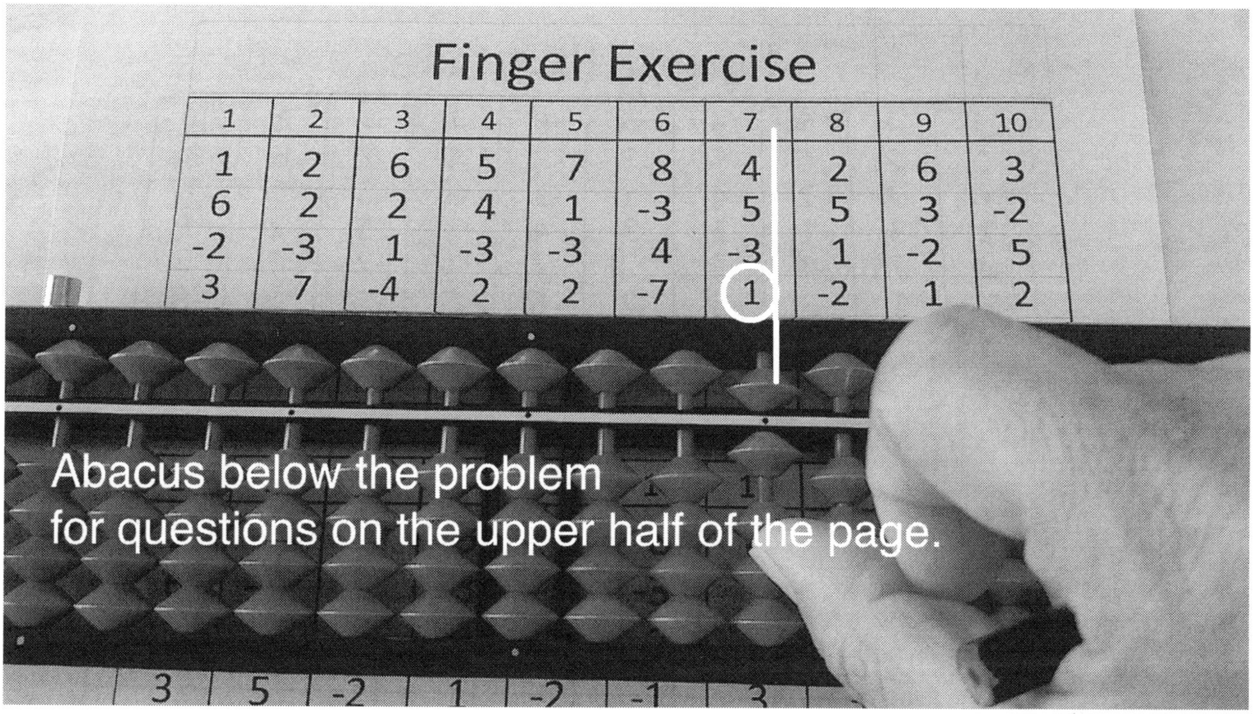
Finger Exercise
1 2 3 4 5 6 7 8 9 10
1 2 6 5 7 8 4 2 6 3
6 2 2 4 1 -3 5 5 3 -2
-2 -3 1 -3 -3 4 -3 1 -2 5
3 7 -4 2 2 -7 1 -2 1 2
Abacus below the problem
for questions on the upper half of the page.

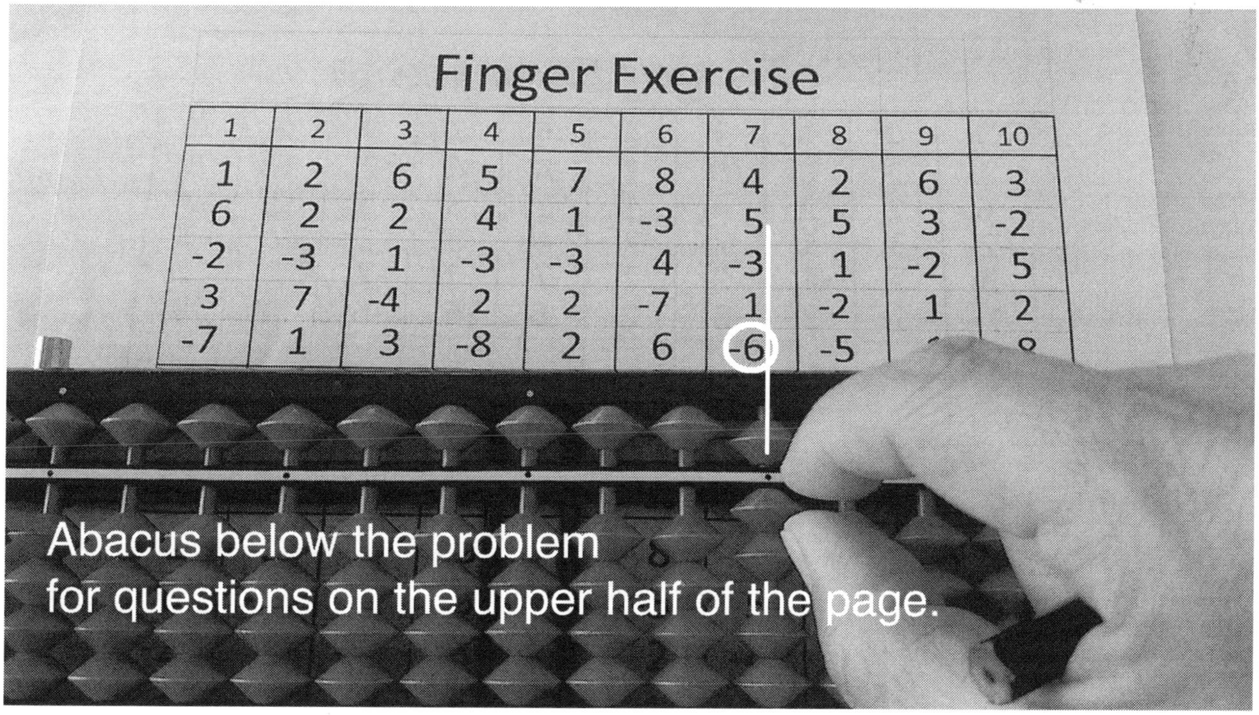
Finger Exercise
1 2 3 4 5 6 7 8 9 10
1 2 6 5 7 8 4 2 6 3
6 2 2 4 1 -3 5 5 3 -2
-2 -3 1 -3 -3 4 -3 1 -2 5
3 7 -4 2 2 -7 1 -2 1 2
-7 1 3 -8 2 6 -6 -5
Abacus below the problem
for questions on the upper half of the page.

Shift down Abacus to read the next number

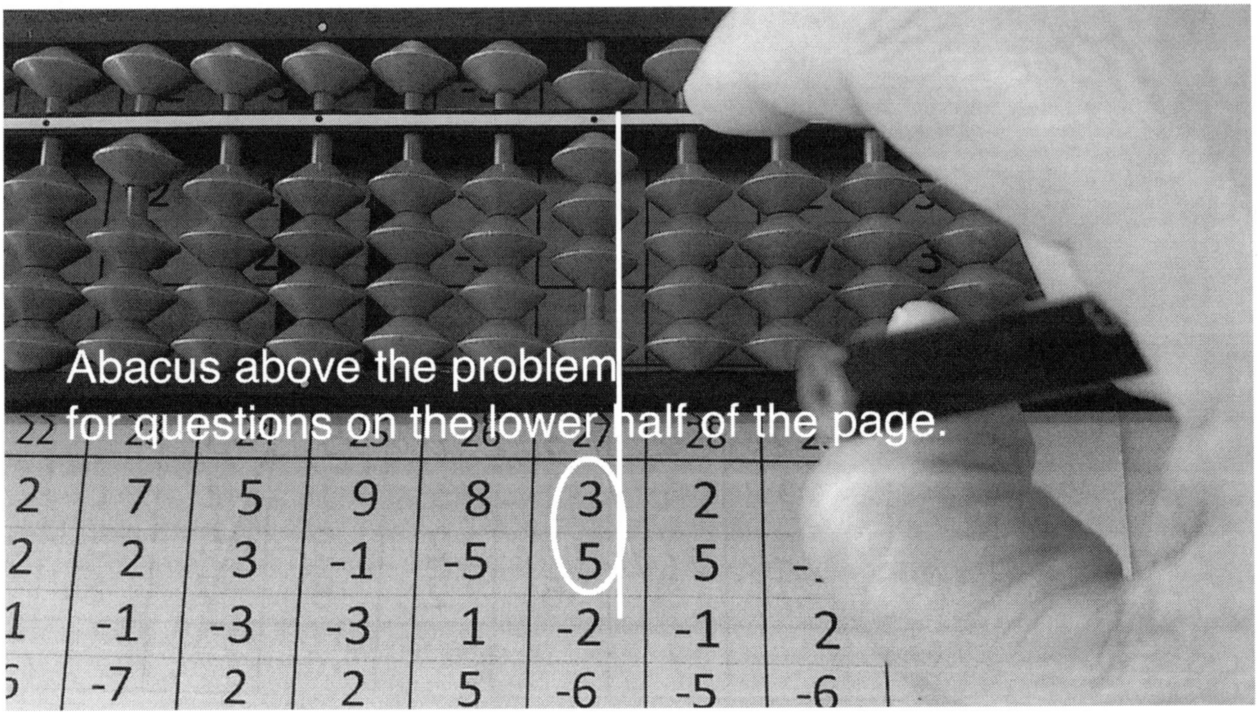

Abacus above the problem
for questions on the lower half of the page.

Abacus above the problem
for questions on the lower half of the page.

Use Abacus to do problems.
If no minus sign is present, that means to add.

Thumb Exercise

1	2	3	4	5	6	7	8	9	10
2 1	1 1	3 1	2 2	4 0	1 2	3 0	2 0	1 3	1 0

Index Finger Exercise

1	2	3	4	5	6	7	8	9	10
5 0	4 -1	3 -1	2 -2	4 -2	3 -2	4 -3	1 -1	5 -5	4 -4

Thumb & Index Finger Exercise

1	2	3	4	5	6	7	8	9	10
6 -1	7 1	1 6	2 7	8 -5	1 8	3 6	2 6	8 -3	9 -4

Use Abacus to do problems.
If no minus sign is present, that means to add.

Finger Exercise

1	2	3	4	5	6	7	8	9	10
1	2	6	5	7	8	4	2	6	3
6	2	2	4	1	-3	5	5	3	-2
-2	-3	1	-3	-3	4	-3	1	-2	5

11	12	13	14	15	16	17	18	19	20
2	2	7	5	7	8	4	2	6	2
6	-1	2	3	-2	-3	5	5	3	2
-2	3	-1	-3	3	2	-4	-1	-5	-3

21	22	23	24	25	26	27	28	29	30
9	2	7	5	9	8	3	2	6	3
-6	2	2	3	-1	-5	5	5	-1	-1
-2	-1	-1	-3	-3	1	-2	-1	2	5

Finger Exercise

1	2	3	4	5	6	7	8	9	10
1	2	6	5	7	8	4	2	6	3
6	2	2	4	1	-3	5	5	3	-2
-2	-3	1	-3	-3	4	-3	1	-2	5
3	7	-4	2	2	-7	1	-2	1	2

11	12	13	14	15	16	17	18	19	20
2	2	7	5	7	8	4	2	6	2
6	-1	2	3	-2	-3	5	5	3	2
-2	3	-1	-3	3	2	-4	-1	-5	-3
3	5	-2	1	-2	-1	3	3	-2	5

21	22	23	24	25	26	27	28	29	30
9	2	7	5	9	8	3	2	6	3
-6	2	2	3	-1	-5	5	5	-1	-1
-2	-1	-1	-3	-3	1	-2	-1	2	5
7	6	-7	2	2	5	-6	-5	-6	-7

Finger Exercise

1	2	3	4	5	6	7	8	9	10
1	2	6	5	7	8	4	2	6	3
6	2	2	4	1	-3	5	5	3	-2
-2	-3	1	-3	-3	4	-3	1	-2	5
3	7	-4	2	2	-7	1	-2	1	2
-7	1	3	-8	2	6	-6	-5	1	-8

11	12	13	14	15	16	17	18	19	20
2	2	7	5	7	8	4	2	6	2
6	-1	2	3	-2	-3	5	5	3	2
-2	3	-1	-3	3	2	-4	-1	-5	-3
3	5	-2	1	-2	-1	3	3	-2	5
-8	-7	3	2	2	-5	-6	-8	7	3

21	22	23	24	25	26	27	28	29	30
9	2	7	5	9	8	3	2	6	3
-6	2	2	3	-1	-5	5	5	-1	-1
-2	-1	-1	-3	-3	1	-2	-1	2	5
7	6	-7	2	2	5	-6	-5	-6	-7
-6	-7	3	-6	2	-8	7	8	6	9

Yay........

You MADE it!

Let's move on to How to Abacus 2 : All About Formulas.

Answer Key

p. 8

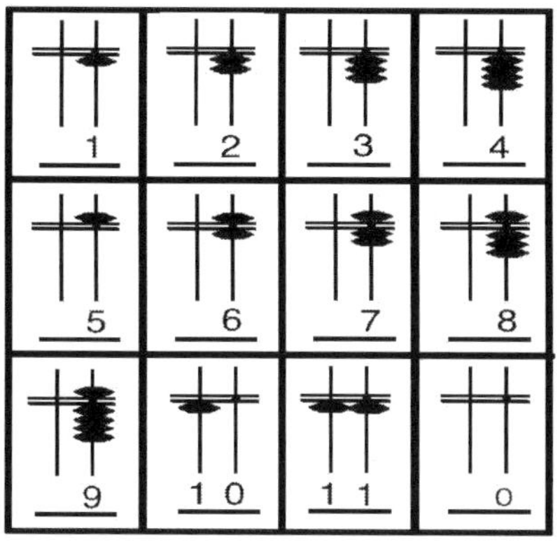

p. 9

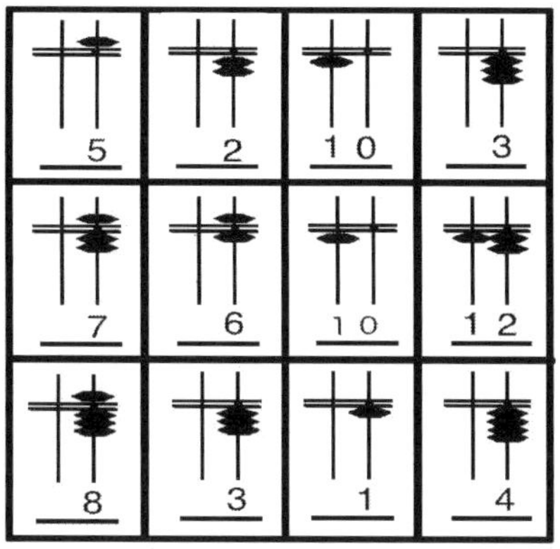

p.10

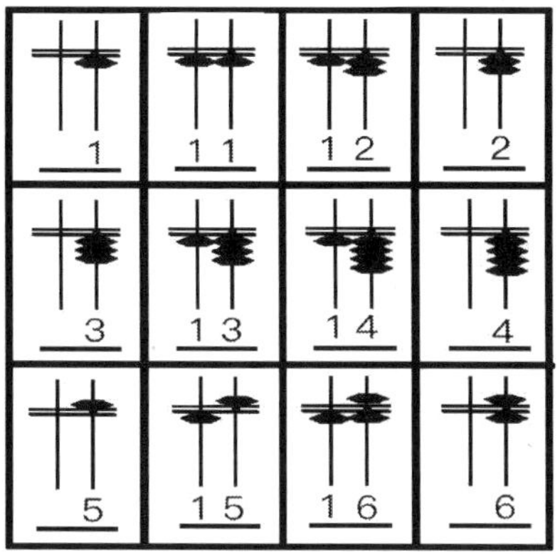

p.11

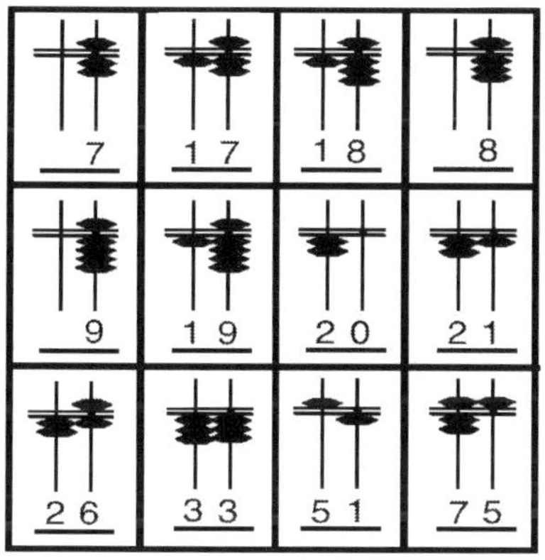

p.12

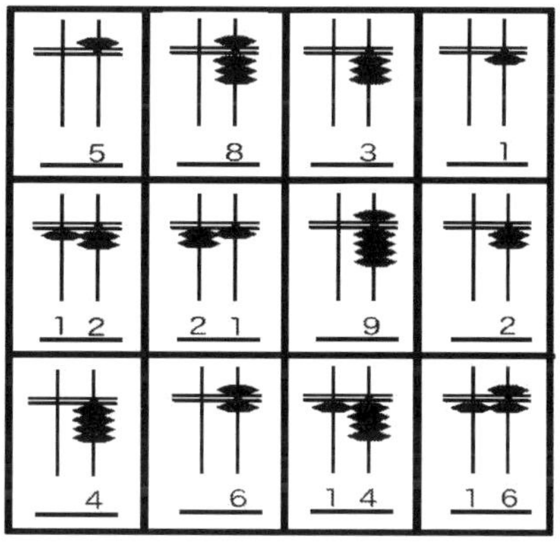

p.13

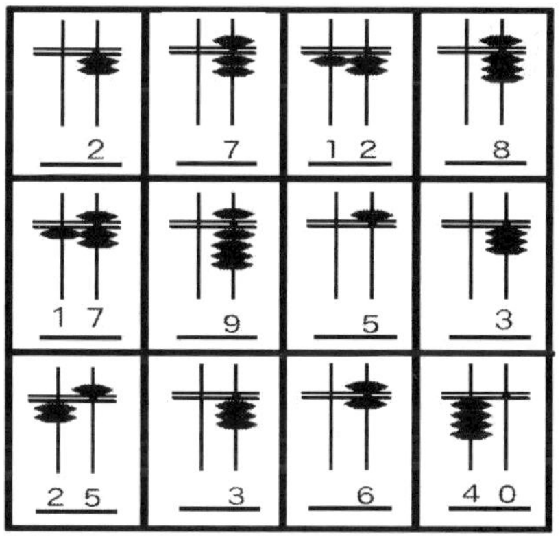

p.40

Thumb Exercise

1	2	3	4	5	6	7	8	9	10
3	**2**	**4**	**4**	**4**	**3**	**3**	**2**	**4**	**1**

Index Finger Exercise

1	2	3	4	5	6	7	8	9	10
5	**3**	**2**	**0**	**2**	**1**	**1**	**0**	**0**	**0**

Thumb and Index Finger Exercise

1	2	3	4	5	6	7	8	9	10
5	**8**	**7**	**9**	**3**	**9**	**9**	**8**	**5**	**5**

p.41, Finger Exercise

1	2	3	4	5	6	7	8	9	10
5	**1**	**9**	**6**	**5**	**9**	**6**	**8**	**7**	**6**
11	12	13	14	15	16	17	18	19	20
6	**4**	**8**	**5**	**8**	**7**	**5**	**6**	**4**	**1**
21	22	23	24	25	26	27	28	29	30
1	**3**	**8**	**5**	**5**	**4**	**6**	**6**	**7**	**7**

p. 42 Finger Exercise

1	2	3	4	5	6	7	8	9	10
8	**8**	**5**	**8**	**7**	**2**	**7**	**6**	**8**	**8**
11	12	13	14	15	16	17	18	19	20
9	**9**	**6**	**6**	**6**	**6**	**8**	**9**	**2**	**6**
21	22	23	24	25	26	27	28	29	30
8	**9**	**1**	**7**	**7**	**9**	**0**	**1**	**1**	**0**

p. 43 Finger Exercise

1	2	3	4	5	6	7	8	9	10
1	**9**	**8**	**0**	**9**	**8**	**1**	**1**	**9**	**0**
11	12	13	14	15	16	17	18	19	20
1	**2**	**9**	**8**	**8**	**1**	**2**	**1**	**9**	**9**
21	22	23	24	25	26	27	28	29	30
2	**2**	**4**	**1**	**9**	**1**	**7**	**9**	**7**	**9**

Made in United States
Troutdale, OR
11/03/2026

56844431R00027